TRIUMPH 2000 AND 2.5PI
THE COMPLETE STORY

Triumph 2000 and 2.5PI

and 2.5PI

◆ *The Complete Story* ◆

Graham Robson

CROWOOD
AutoClassic

First published in 1995 by
The Crowood Press Ltd
Ramsbury, Marlborough
Wiltshire SN8 2HR

British Library Cataloguing in Publication Data
A catalogue record for this book is available from the British
Library.

ISBN 1 85223 854 2

Picture Credits
Photographs kindly supplied by Mirco Decet, Gordon Birtwistle,
Roy Fidler, Alan Taylor, John Hopwood, Peter Browning and
Triumph Sports Six Club.

Typeset by Phoenix Typesetting, Ilkley, West Yorkshire.
Printed and bound in Great Britain at The Bath Press.

Contents

Acknowledgements

It has taken many years, many interviews, and a lot of generous help from busy people for me to be able to compile this book. I salute all the one-time directors, managers, engineers and works drivers of Standard-Triumph for their help. What a pity it is that some of them are no longer with us to see how the 2000s and 2.5PIs are cherished in the 1990s.

To separate the facts from the fantasies, to listen to all the best stories, and to get to the bottom of all the events which influenced the career of these cars, over the years I have talked to: Ray Bates, Gordon Birtwistle, Bill Bradley, Peter Browning, Frank Callaby, Harry Colley, Brian Culcheth, Alick Dick, Alan Edis, David Eley, Roy Fidler, Ray Henderson, Spen King, George Jones, John Lloyd, Maurice Lovatt, Bill Price, John Sprinzel, Alan Taylor and Harry Webster.

Then there are the many experts on matters historical, starting with members of the British Motor Industry Heritage Trust – David Bishop, Anders Clausager, Peter Mitchell – and Michael Ware and his colleagues at the National Motor Museum; I should also like to add my thanks to all the British motoring magazines of record, whether contemporary (from the 1960s and 1970s) or those of the 1990s specializing on 'classic' cars. Plus, of course, Richard Langworth, my personal North American connection, who always manages to dig up more facts on Standard-Triumph in that continent when I need them.

Thanks also to Mike Jackson of the Triumph 2000/2.5 Register for helping to reunite Roy Fidler and Alan Taylor with their famous old 2000/2.5 Prototype rally car, and to Mirco Decet for taking so many splendid photographs.

Most of all, of course, my thanks go to all those enthusiasts who enjoy owning, driving and preserving so many of the cars.

Introduction

I have always thought that the Triumph 2000 was an underestimated car, so I am delighted to be able to tell what I hope is the complete story of its career. Most people would agree, I am sure, that this was one of the most elegant cars ever to carry the Triumph badge – and it was also one of the most successful.

It was a typical Standard-Triumph exercise of the period: versatile, profitable and well-regarded. From one basic design were derived saloons and estates, engines with carburettors and fuel injection, and two distinctly different styles. Fastback types were considered, and an attractive Grand Tourer (the Stag) also evolved from the same basic layout.

But I had better start with a confession. I intend to bend one of the ground rules of the historian's craft: I am not going to be completely impartial, and I freely admit it. In the 1960s I was often too close to the cars, to their development, and particularly to their competition programme, for that. What follows, therefore, is not only a book of record, but one that includes a mass of personal reminiscence, some of it mine, but much of it from personalities who helped make the cars successful.

Between 1963 and 1977 well over 300,000 cars of this type were built, and it is a measure of their strength and their excellent engineering that a good many survive to this day. Throughout this fifteen-year career they not only recovered a reputation the last of the Standard Vanguards had lost, but they fought head-to-head, and successfully, against another new-generation 2000 (the Rover), and more extensively backed models from BMC, Ford and Vauxhall.

The downside is that Standard-Triumph took ages to settle on the new design. For the very first time, I believe, this book tells the tortuous story of an evolution that began with the Standard Vanguard Six, which stumbled agonizingly through various Zebu layouts, and finally settled on Barb.

The rest, though, is history, fascinating, colourful history, which I hope all enthusiasts will enjoy.

7

TRIUMPH 2000 AND 2.5PI EVOLUTION
NAMES, DATES, PLACES AND INFLUENCES

Although the original Triumph 2000 was launched in 1963, and was finally dropped in 1977, many other events influenced its career.

Before Leyland 1948–60

1948	Original Standard Vanguard put on sale.
August 1957	Zebu project, a six-cylinder-engined saloon, first discussed at board level.
September 1960	Zebu project officially dropped.
	Re-engineered American Motors Rambler briefly considered in its place.

From Barb to 2000 1960–8

December 1960	*Leyland Motors launched a take-over bid for Standard-Triumph. This was formalized in May 1961.*
Spring 1961	Michelotti started work on Barb – which became the 2000.
November 1963	Final approval of Barb style given.
May 1963	*The last Standard-badged private car, an Ensign De Luxe, was assembled.*
October 1963	Triumph 2000 saloon introduced. First series-production cars delivered in January 1964.
October 1965	Triumph 2000 Estate introduced.
December 1966	*Leyland announced agreed take-over of Rover, whose 2000 was a direct rival to the Triumph 2000.*
January 1968	*British Leyland founded by amalgamating Leyland with BMH.*

The 2.5PI 1968–75

October 1968	2.5PI derivative of 2000 announced.
October 1969	Restyled Mk II versions of 2000/2.5PI announced.
June 1970	New Stag 2+2 Cabriolet model, based on 2000 design, introduced.
February 1972	*Styling of new Rover SD1 approved. This range eventually replaced both the Triumph 2000 and Rover P6 models.*
March 1972	*Rover–Triumph set up as division of British Leyland.*

Final Modifications 1974–7

May 1974	Introduction of 2500TC models.
April 1975	2.5PI models discontinued.
July 1975	Introduction of 2500S models, and revision of existing 2000 (became 2000TC) and 2500TC models.
July 1976	*New-generation Rover SD1 model introduced.*
May 1977	Last 2000/2500 models assembled.
June 1977	Last Stag model built.
October 1977	*Introduction of Rover 2300/2600 models.*

1 Ancestors – Standard, Triumph and Vanguard

Where does the story of a famous car really start? When it first went on sale? When it was conceived? Or when its own ancestors were unveiled? In the case of the Triumph 2000, so many other influences – people, cars and companies – helped to shape the car, that I was happy to go back a long way. In fact, to tell *all* the story I needed to start at the end of 1944, when Standard took over the Triumph name.

Although the Triumph 2000 was revealed in October 1963, this story really began many years earlier in the 1940s, when the world was still at war, when there was very little frivolity in evidence, and when fashion and status took a back seat. The Standard Motor Company was one of Coventry's largest enterprises, while Triumph, once an important car maker, was moribund.

At this time Standard's factories were building military aircraft (including the famous Mosquito fighter-bomber), aero engines and armoured cars, while the single remaining Triumph factory that had not been flattened by enemy bombing (the 10¾ acre (4 hectare) Gloria Works) had been sold to the Air Ministry, and was producing aircraft carburettors and many other military components. There was frantic activity round the clock seven days a week. Almost everyone worked under great pressure, and few had time to look ahead and plan for a peaceful future.

Obviously, Triumph in 1944 was already vastly different from Triumph in the 1930s. The original Triumph company, which had begun building motorcycles in 1902 and followed up with cars in 1923, hived off the motorcycle-making side of the business in the mid-1930s. The car-making concern, the Triumph Company Ltd, suffered a terminal financial crisis in 1939, after which its new owners (Thos W. Ward Ltd, of Sheffield) made no plans to start building cars again.

The Standard Motor Co. Ltd, on the other hand, had started building cars in 1903, prospered ever after that, boomed remarkably in the 1930s with Captain John Black running the business, and was running a vast multi-factory business by 1939.

By 1944, and in spite of the demands on his time, Standard's managing director, Sir John Black as he had become, already had grandiose post-war plans. Not only did he send his chairman (Charles Band) and his young assistant Alick Dick to buy the remains of Triumph from their temporary owners, but he tried to attract Cecil Kimber (one-time guiding genius of MG) to join him to run a revitalized company. Added to this (for his hunger for power and prestige knew no bounds), he also entertained the preposterous idea of taking over Jaguar Cars, but its sole owner and founder, William Lyons, would have none of that. As Alick Dick once told me :

Captain John Black (right) congratulates Standard's technical director Ted Grinham (left) on the birth of the Standard Flying Nine in 1936. Both were still directors of Standard in the 1950s, the old guard who made sure that innovation was slow and styling basically unadventurous.

Just before the end of the war, in 1944, Triumph itself was effectively still bust. John Black wanted another name alongside Standard – just that. As far as I can remember, Sir John sent me out to the [3½ acre (1.5 hectare) Triumph] Stoke works with Charles Band, Standard's chairman. We looked at the place, which wasn't worth a farthing. But we went ahead and bought the place . . .

A farthing? This was still a coin of the realm at the time – one quarter of an old penny!

Triumph's more modern factory in Foleshill, having been sold off already, was not even part of the deal. All that was on offer was the blitzed remains of the older Briton Road factory, and various rights. On 20 October 1944 Standard's board of directors confirmed a plan to buy what was left of the Triumph business. For a total of £75,000, Standard gained the freehold of a site, any and all jigs, tools, parts and drawings which remained and '. . . the goodwill of the business, with the right to use the name Triumph in conjunction with any company to be registered by the Standard Motor Co. Ltd . . .'

Sir John Black, it should be emphasized, only wanted the name, and had no intention of reviving the pre-war designs. Alick Dick recalled that:

We sold the factory to the B.O. Morris Group, and all we got was the name – for nothing – plus, I suppose, an obligation to supply spares for Triumph cars. But as there were hardly any of them, we really didn't feel any obligation to the old customers.

Sir John had no intention of rebuilding the old Triumph company. As Alick Dick confirms:

Standard had been supplying parts to SS. Standard had special tooling for the six-cylinder engines, and for the overhead-valve version of the four-cylinder, of which Lyons took all supplies. Bill Lyons wanted to buy the tooling – he wanted to make all of his Jaguars in the future – and John Black was willing to let him have that so that we could build a competitive car. We kept the four-cylinder tooling, but it just wasn't viable without a new chassis and a new name. So that's why we bought Triumph. Just because Bill Lyons made a sporting saloon, or a sports car, John Black was not going to let him get away with it!

Even so, Triumph's progress to glory would not be swift, easy or, indeed, logical. What John Black had in mind for his new purchase in 1944 was not the way that Triumph became lastingly famous in the 1960s . . .

Once started by Standard, the absorption of Triumph into the larger company was swift and final. The formal conveyance of Triumph to Standard's control was dated 24 November 1944, a new Standard subsidiary called the Triumph Motor Company (1945) Ltd was set up soon afterwards, holding its first board meeting in April 1945.

STANDARD AND TRIUMPH

Was there a strategy at first? If there was, it changed dramatically in the next few years. First, Sir John Black failed to attract Cecil Kimber to work for him. After that, any belief that his design team could produce viable competition to the current SS-Jaguars was swiftly shattered. His fall-back position was to use the 'Triumph' badge on a series of up-market (and more profitable) Standards, and even that was a shaky strategy at first.

Four months after the take-over, the public learned a little more. Early in 1945 Standard issued a press statement which confirmed that: 'With the object of getting back into their production stride as rapidly as possible after the cessation of hostilities the Standard Motor Company Ltd has decided to concentrate on two up-to-date models, an Eight of 1,000cc and a Twelve of 1,600cc. *It is also proposed to produce a 10hp Triumph of 1,300cc and a 15hp model of 1,800cc*' [author's italics].

If only we had realized it, this was evidence that Standard was already thinking of building cars around Standard's engines – the '10hp' machine would appear in 1949, badged as a Triumph Mayflower, while the '15hp' model would appear in 1946.

Although no mention was made of Sir John's truly major post-war plans – which were to develop an all-new, medium-size Standard family car, and to find civilian ways of using the vast Banner Lane 'shadow' factory that Standard had been running for the British government – it was clear that his revised master plan for Triumph was already developing. Standard would produce the bread-and-butter models, while cars badged as Triumphs would provide the glamour, the gloss and, he hoped, enhance the company's image.

The First Standard-Triumphs

This short section illustrates that great men – even great tycoons, of whom Sir John Black was a perfect example – often flounder when trying to move their companies into new territory. It took Standard nearly a decade to develop a proper sporting image for Triumph. The breakthrough came with the launch of the TR2 sports car, the arrival of the Herald was a great boost, and it was the subject of this book, the Triumph 2000, which eventually killed off the 'Standard' marque completely.

The original 'Standard-Triumphs' were the 1800 Roadster and the 1800 Town and Country Saloon. These were launched in 1946, using the same type of tubular chassis frames with different wheelbase lengths, but with engines, transmissions, axles and suspensions lifted straight out of the existing (1930s) Standard Flying Fourteen family cars. The Roadster had a Standard-produced body (panels were produced on aircraft-type rubber presses that Standard had used during the Second World War), complete with a

The original Standard product to be badged as a Triumph was the Roadster, built from 1946 to 1949. There was no overlap with independent Triumphs of the 1930s, for this car used Standard Flying Fourteen engines, transmissions and suspensions.

The Triumph 1800 saloon of 1946 used the same basic chassis as the Roadster, but in longer-wheelbase form, so this car was really a Standard Flying Fourteen 'in a party frock'. The razor-edged styling was a throwback to fashionable shapes of the 1940s.

The Vanguard saloon of 1948 was Standard's all-new post-war car, which aped early-1940s styling, putting strength and a roomy interior ahead of roadholding and character. Not an inspiring machine.

dickey seat, while the Saloon had a razor-edge style of saloon produced by Mulliners of Birmingham.

From 1949 both these cars were given Standard Vanguard running gear, and the saloon was eventually badged as a Renown, but they were always expensive and sales were somewhat disappointing. Between 1946 and 1954 only 4,601 Roadsters and 15,491 Saloons were produced.

By the mid-1950s there were plans for a Renown replacement, but happily a scheme to badge a more powerful version of the Vanguard Series III as a Triumph Renown was dropped at the last minute. Instead, Standard named this car the Vanguard Sportsman, which was an awful machine of which less than 1,000 were ever sold.

The next move was altogether more ambitious, for although the Mayflower of 1949/1953 used a modified Standard Ten engine and Vanguard transmission components, it had a unique monocoque body

In 1953 Standard reshaped the Vanguard, but it was still a lumpy machine which ran on a very uninspired chassis. The engine, however, was the famous wet-liner four-cylinder unit that proved to be so useful in the Triumph TR sports cars.

The Triumph Mayflower, on sale from 1950 to 1953, was another Standard effort to sell an up-market saloon model. It was misguided, not only because of its strange and obsolete razor-edge style, but by its use of a side-valve 1.25-litre engine.

style, and independent front suspension. The short wheelbase combined badly with the chosen type of razor-edge styling; this, combined with relatively high prices, ensured that only 35,000 were produced.

Triumph TRs and Heralds – exciting new-generation models

Later in the 1950s, Triumph suddenly became successful and fashionable, with cars as smart and sporting as the recent Standards had been dowdy and uninspiring. Even though the TR2 sports car of 1953 used a mixture of further-developed Vanguard and Mayflower chassis components, it was a fast, economical and rugged little two-seater, which rapidly built up an image. Within five years, and starting from scratch, Triumph was already fighting head-to-head for supremacy against Austin-Healey and MG in the North American market.

Then came the Triumph Herald family of 1959, not only a new Triumph to replace the Standard Tens and Pennants, but one that had been excitingly styled by the little Italian genius, Giovanni Michelotti, with real flair.

To see the transformation under way at Canley, therefore, one only had to compare the Standards of 1954 with the Triumphs of 1959. In every way, it seemed, there had been a completely new approach in style, in character, in performance and in badging. The Standards had looked, behaved, and were, lumpy and boring cars without an ounce of fun in their make-up, the Triumphs that followed looked good, performed well, and encouraged their owners to go out and enjoy themselves.

A New Team

But how? Quite simply, none of this would have been possible without a total modernization of the management. It was not merely that the old autocrat, Sir John Black, had been ousted by Alick Dick, but that most of his cronies had soon followed him. Technical director Ted Grinham, whose ideas were rooted in the 1930s, had given way to Harry Webster; stylist Walter Belgrove had gone, replaced by Giovanni Michelotti, while other thrusting characters such as Martin Tustin and George Turnbull had also burst on to the scene.

When I look back at the characters who shaped Standard in the 1940s and early 1950s it is easy to see why the cars were being built that way. A few simple 'before and after' pen portraits make this point.

In the early 1950s Standard was managed by a group of old men. Sir John Black, who had been unchallenged at the helm of Standard since 1934, was a ruthless dictator, fifty-eight years old when he was finally pushed out of the company in January 1954, and a man succinctly described by Alick Dick (who knew him better than any other Standard-Triumph executive at the time) as a fully paid-up schizophrenic who was getting worse all the time.

For many years Sir John's chairman had been Charles Band, a Coventry solicitor whose first links with Standard had been forged in 1916 (he helped negotiate the purchase of the land on which the Canley factory complex was originally developed). Band's main interest was in finance, certainly not in the product. When he retired, in 1953, he was seventy-nine years old.

Ted Grinham, a dour and humourless man, had been Standard's chief engineer from 1931 and its technical director from 1936, but his work, though sound, never showed the slightest flair or imagination. Happy to bow to Sir John's wishes, if only to achieve a quiet life, he was happy to perpetuate side-valve engines, three-speed gearboxes with steering column changes, and bench front seats, for he could see little merit in designing fast cars, and never drove quickly himself. Harry Webster, who worked under him for many years, barely remembers an occasion when he smiled, unless the press cameras were pointed at him.

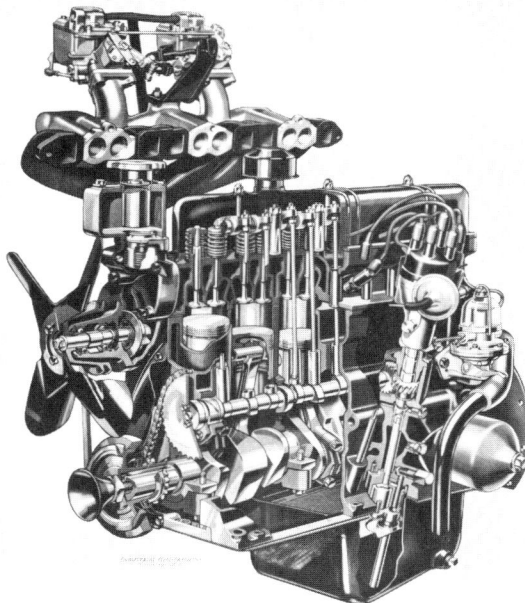

The first of the building blocks that would help make up the Triumph 2000 was the six-cylinder engine. Designed in the 1950s, and first used in the Standard Vanguard Six of 1960, it was a sturdy, no-nonsense unit which proved to be extremely versatile. In its Vanguard Six application, it was fuelled by two semi-downdraught Solex carburettors.

Walter Belgrove, who had been Triumph's chief body engineer in the 1930s, took up that post at Standard immediately after the war. In the next decade Belgrove seems to have been perpetually in dispute with Ted Grinham, always in awe of Sir John Black, and in later life seemed to carry so many chips on his shoulder that there was little space for anything else. Some of his styles were good, but some – like the original Vanguards, and the Standard Eight/Ten family – were dumpy in the extreme. After

Sir John Black (extreme right) and his assistant Alick Dick inspecting the first Mayflower production shell to go down the assembly lines at Canley in 1950.

yet another blazing row with Ted Grinham he stormed out of the company in 1955.

The team that took over in the mid-1950s was younger, more vigorous, and more interested in cars as machines to be enjoyed. With the single exception of the chairman, Lord Tedder, they were a whole generation younger.

Alick Dick, the son of a Norfolk country doctor, was perfect 'Hollywood casting' for the job he took over in 1954. Once an apprentice at Standard, later Sir John Black's personal assistant, then assistant managing director from 1949, he had come up through the ranks in a mere two decades. He was only thirty-seven years old when he became managing director in 1954. By that time he already had a glittering reputation in the motor industry, had flirted briefly with rallying in his youth, had a penchant for round-the-world business trips, and as a handsome, always-smiling company figure-head was often in the news. When Leyland Motors took over in 1961, he rapidly fell out of favour: as author Graham Turner once observed, he was a Cavalier suddenly cast among Roundheads.

From June 1954 Dick's chairman was Marshal of the Royal Air Force, Lord Tedder, the Second World War hero who had run the RAF effort in the Middle East from 1941 to 1943, and had been General Eisenhower's deputy in SHAEF. Although Tedder knew little about cars, at least he came from a gallant profession, and had a great reputation with the public. In fairness, though, I should emphasize that he was more of a figure-head, placed in the chair to impress the City of London, than a practising motor industry character. When he joined Standard in 1954, he was already nearly sixty-four, and he retired at the age of seventy, in 1960.

Harry Webster, who soon became Dick's chief engineer, had also started as a

---

Alick Dick (managing director of Standard-Triumph, 1954–61)

Under Alick Dick's direction, the marketing balance at Canley tilted emphatically from Standard to Triumph at the end of the 1950s. By the time he left the company, ousted by his new masters at Leyland Motors, it was inevitable that the replacement for the Standard Vanguard would be called a Triumph.

In seven turbulent years while he was at the helm, the company changed from a staid organization with boring products into one that embraced sport and modern design. There was, quite literally, no comparison between the Triumph Herald and the Standard Eight and – as we would see in 1963 – none between the Triumph 2000 and the Standard Vanguard.

Until he left Standard-Triumph in September 1961, Alick Dick had been loyal to the company throughout his working life. The son of a Norfolk country doctor, he had joined Standard as an apprentice in 1934, became Sir John Black's personal assistant in 1947, and deputy managing director in 1951. He led the boardroom revolt that deposed the autocratic Sir John in 1954 and succeeded him, at which point he was still only thirty-seven years old.

The 'young lion' (as Britain's national press liked to describe him), instituted an utterly different management style from that of Sir John Black, and later told me that among his colleagues he was merely the *primus inter pares* (first among equals). While the Herald was being developed, and the work leading towards the birth of the 2000 was going ahead, he relied heavily on the expertise of three particular people: Harry Webster as his technical chief, Martin Tustin as his general manager, and George Turnbull as his production specialist.

During the 1950s he thought it vital for Standard to find a big industrial partner, and if any of the merger talks he had with the Rootes Group and with Rover had come to fruition, an earlier version of the Triumph 2000 might have appeared. Frustrated by a lack of success, and by growing disagreements with Massey Ferguson (for whom Standard built tens of thousands of tractors), he decided to make a dash for individual growth, buying up body makers like Mulliners of Birmingham, suspension specialists like Alford & Alder of Hemel Hempstead, and planned to build a vast new factory at Speke, on Merseyside.

In the meantime, project work on Zebu (as the Vanguard-replacement was coded) staggered from crisis to crisis, while as a result of a government credit squeeze there was an abrupt turndown in the British market in 1960. Profits turned into losses, work on the Zebu project had to be cancelled, and Dick was forced to accept a take-over bid from Leyland Motors to ensure Standard's survival.

Leyland formally took control in the spring of 1961, by which time styling and project work on Barb (the definitive Triumph 2000) had begun. In September 1961, well before Dick and his fellow directors were forced to resign, the Triumph 2000 had been designed.

Alick Dick died in March 1986, less than two years after the last Triumph-badged car of all (the Acclaim) was produced. He was only sixty-nine years old.

Standard apprentice in the 1930s. As chief chassis engineer from 1949, he inspired the birth of the TR2 and the Herald. When Alick Dick took the helm in 1954, Webster was the same age – thirty-seven.

The personality who completed this dynamic team was an energetic young Italian stylist, Giovanni Michelotti.

Although he did not even become involved with Standard-Triumph until 1957 – three years after Alick Dick had taken command – when he was still only thirty-six, his designs had already appeared on many exotic Italian supercars. From 1957 until the early 1970s, Michelotti inspired the shape of almost every new Standard and Triumph model.

Replacing the Vanguard III

Standard's original Vanguard had been launched in 1947, a tough, no-nonsense design with the same four-cylinder engine that Standard had designed for use in the Ferguson tractor. The original cars had reliability, long life and simple design but precious little flair. Ten years on, the third-generation model, the Vanguard III, looked well but was no more refined than before.

In 1957, only two years after the mono-coque Vanguard III had been put on sale, Standard-Triumph's management started thinking about a successor. As far as the sales force was concerned it could not come soon enough, for they knew, full well, that the Vanguard era was over.

By comparison with the latest six-cylinder cars from Ford, Vauxhall and BMC, the four-cylinder engined Vanguard was no longer refined enough. As Harry Webster made clear, when talking about 'Zebu', the first attempt to replace the Vanguard, the old wet-liner engine was never smooth. Other engineers recall that: 'We could no longer compete with the opposition with that big four-cylinder engine in the Vanguard.'

Even more important was that the Vanguard's road manners and stability, particularly on rough roads, left a lot to be desired. As development engineer Maurice Lovatt told me: 'When the first-ever comparisons could be made on pavé tracks between a Vanguard and an all-independent 2000 prototype, the new car was almost unbelievably better.'

Even though Michelotti's genius produced the Vignale Vanguard, to freshen up the style, and the new 20S six-cylinder engine was finally made available in 1960, the end could not be long delayed.

Even so, design of the new car would have to take its place in the queue of priorities, for the department was small, and there was always more work than the capacity to get it done. The layout of the Zobo family saloon (which became the Herald) took top priority until 1958, and the development of the TR4 sports car also got in the way after that. Starting again, after the Vanguard, would take some time.

2　The Zebu Project

Although the Triumph 2000 was not revealed until 1963, the first moves to replace its predecessor, the Standard Vanguard III, were made in 1957. By that time the monocoque Vanguard had been on sale for two years, and except for the plan to fit the new SC-based six-cylinder engine in place of the old wet-liner 'four', there were no plans to make any further major changes to that car during its life.

Even though the four-cylinder engine had established a great reputation for reliability, it was never a refined unit. Harry Webster himself still thinks that 'big fours' over 2-litres could never be made smooth enough – this period, of course, being years before Lanchester-type balancer shafts were re-invented. As far as he, and all the directors were concerned, any new car would *have* to have a 'six'; it was necessary, in any case, to compete with BMC, Ford and Vauxhall, who already had six-cylinder engines on the market.

By then, as I have already made clear, the old guard had retired, and a new, young management team, led by Alick Dick, had started to look ahead. While Dick looked around the industry for a partner to help him secure Standard's long-term future, his engineers (led by Harry Webster) began to develop several new model families.

What had looked like a straightforward task in 1954 got progressively more complicated in the years that followed. In six years Alick Dick discussed mergers or major co-operation with Rover (twice), Dennis Brothers, Rootes, Massey-Harris-Ferguson and Leyland, and in the same period he also bought up smaller companies such as

Mulliners, Hall Engineering and Alforder Newton to help him expand the company's base.

'My mission at Standard-Triumph,' Dick told me, 'was to look around for a partner, a partner to secure our future. In the next few years I had flirtations with almost every other independent concern except Chrysler. I never talked to them.'

He also fought off a take-over bid from Massey-Harris-Ferguson, as a result of which he decided to divorce Standard-Triumph from the tractor-building interests of that company. Standard came out of that fracas with £12 million cash, and commissioned a huge expansion of the company's assembly and body-making facilities along the way!

Webster's small but incredibly resourceful team faced similar complications. Faced with the difficulty of replacing the Standard Eight/Ten (they had been told that the existing Fisher & Ludlow body-making facilities would not be available to them when that range was dropped), and of settling the style of a sports car to replace the TR3, they had to let the Vanguard replacement project slip into the background for a time. One immediate result was that the six-cylinder engine primarily intended for the Vanguard replacement first had to be installed in the Vanguard itself.

THE PLAN

Even so, by mid-1957 the bare bones of an all-new car were taking shape. Harry Webster, who had been appointed as a special director

Standard-Triumph four-letter code names

In the 1950s and 1960s Standard-Triumph decided to christen each new design project with a code word of four letters. This was not only to preserve some sort of secrecy when originally dealing with suppliers, but the names also became a type of verbal and written 'shorthand' when directors and managers were discussing the future with each other. Harry Webster asked one of his staff, Noel Fielders, to work up the original list, originally using words beginning with the letter Z.

None of the names chosen were meant to have a hidden (or Freudian) meaning but, as Webster reminds us, Zobo 'is a Tibetan pack-animal of indeterminate sex somewhere half-way between a bull and a cow'! Zebu, the project that preceded the Triumph 2000, is defined as a 'humped ox'.

The 2000 itself was Barb – but many other names were used during the period. Here are the most important ones :

Able	Atlas Major commercial vehicle	Wasp	Triumph TR5
Ajax	Front-wheel drive Triumph 1300	Zany	Standard Atlas commercial vehicle
Atom	Triumph Vitesse		
Barb	Triumph 2000	Zarf	Standard Vanguard III pick-up
Beta	Wide-track TR3A, cancelled	Zebu	Six-cylinder project of 1957–60, which preceded the Triumph 2000, cancelled
Brig	Leyland 20 (revised Standard Atlas Major)		
Bomb	Triumph Spitfire	Zero	Standard-Triumph tractor project, a proposed successor to the Ferguson tractor. Cancelled
Fury	Prototype 6-cylinder sports car, with monocoque shell, cancelled		
Lynx	Predecessor to Triumph TR7, name later used for long-wheelbase TR8 hatchback, cancelled	Zest	Definitive TR4, which succeeded the Zoom
		Zeta	A proposed Michelotti restyle of the Standard 8/10, using that car's floor pan. Not to be confused with the Standard Pennant. Cancelled
Manx	Short-tail/cheaper version of Triumph 1300. As Manx II/Manx IV, evolved into Toledo/1500		
Pony	Lightweight 4x4 vehicle, using modified Triumph 1300 engine/transmission units	Zobo	Triumph Herald project
		Zoco	Standard Atlas Major derivatives
		Zoic	Standard Vignale Vanguard (modified SIII)
Puma	Schemes for a replacement for the 2000 range, cancelled in favour of Rover SD1	Zoic II	Standard Vanguard Six
		Zoom	Long-wheelbase pre-TR4 shape, by Michelotti (1960–1 TRX race cars used this basic shape), cancelled
Stag	The car that became the Triumph Stag		
Star	Revised chassis for the Herald/Vitesse		

Other, longer, code names and numbers gradually took over from the late 1960s. The 2000 Mk II, for instance, was always code-named Innsbruck.

When Standard decided to develop a successor to the Standard Vanguard, they started the Zebu project. This was the very first full-size mock-up of Zebu, complete with reverse-slope rear window, as conceived in 1957.

of engineering in May 1957 (succeeding Ted Grinham, who had retired at the end of 1956), was soon ready to brief his colleagues. In August 1957 the directors were told about a new project, coded Zebu, which would follow up the Zobo (Triumph Herald) model through the experimental workshops.

First thoughts were that this car should have a separate chassis, with styling to be carried out by Standard-Triumph, but with advice from Michelotti, and with bodies tooled and supplied by Pressed Steel Co. It was to have the newly developed six-cylinder engine of 1.5 or 2 litres, the transmission was to be at the rear of the vehicle, and there would be all-independent suspension. In fact all the prototypes were powered by 2-litre engines. Pneumatic suspension was being considered (though, as far as I can see, this was never tested on a car), and it was hoped that the first production cars would be built in March 1960.

By that time the company was hoping to have capacity to build 161,000 cars a year, of which 110,000 would be Herald-family and 30,000 would be Zebu-based.

This was the proposed timing plan:

Mid-October 1957	Full-size wooden model available – final styling to be decided.
December 1957/ January 1958	Full-size wooden model sent to Italy for. completion of first prototype body. Delivery of complete chassis to Italy for mounting of first prototype body.
April 1958	Delivery from Italy of the first prototype.
May 1958/ June 1958	Full-size wooden model to Pressed Steel Co. for engineering under Standard Motor Co. supervision.
December 1958	First production prototype due from Pressed Steel Co.
March 1960	Start of production.

The UK target price of Zebu was to be £575 (before Purchase Tax was added) at 1957 levels, which was quite ambitiously low. This compared with the Vanguard III De Luxe's list price of £650. The current Ford Zephyr

21

Six was selling for £610, and the Vauxhall Velox for £580.

Although this was to be an early use of Standard's first modern six-cylinder engine, the engine design itself was by no means new. Based on the same basic layout as that of the small SC four-cylinder engines as used in Standard Tens (and, soon, the Herald), but with much different detailing, it had first been proposed in 1952. Five years on the engine existed, at least in prototype form, but was not yet ready for production.

SC, the all-new small 'four' that powered

Triumph's new 'six' – engine origins

As is well known, the straight six-cylinder engine, which was used in all 2000s, 2500s and 2.5PIs was a direct descendant, and close relative, of the famous Standard-Triumph SC (small car) four-cylinder engine. The SC 'four' was conceived in 1951, and went into production in 1953.

The 20S 'six', though, had a much longer gestation period. Conceived in 1952, first run in the mid 1950s, but not fitted even to a prototype car until 1957, it finally went on sale at the end of 1960. In the years that followed, more and more versions were built, the last of all being produced in 1977.

Bore and stroke dimensions always give a clue to a engine's bloodline. Here are all those that apply to the SC 'four' and the 'six'.

Capacity	Bore x stroke (mm)
SC (4-cylinder)	
803cc	58 x 76
948cc	63 x 76
1,147cc	69.3 x 76
1,296cc	73.7 x 76
1,493cc	73.7 x 87.5
(6-cylinder)	
1,422cc	63 x 76 (first prototype only)
1,596cc	66.75 x 76
1,998cc	74.7 x 76
2,498cc	74.7 x 95

---

the Standard Eight of 1953, was designed in 1951, and at first much of the plant used to manufacture the components took over machinery and factory space previously allocated to Triumph Mayflower facilities; this explains why the SC's cylinder centre positions were identical with those of the side-valve Mayflower engine.

In the long term, however, the SC family was always meant to be built on new transfer machinery, which actually came into use during 1955. Not only that, but the six-cylinder version was once described to me as 'really the small "four" with two more cylinders stuck on the end. Both units looked the same at the front and rear of their cylinder blocks, but had six instead of four holes in the middle . . .'. Where possible, it was always intended to be machined on some of the same transfer tooling. Standard was well-versed in this technique, for the company had first started doing it as early as the 1930s, with four-cylinder and six-cylinder versions of the famous 106mm stroke engines used in many Flying Standards, plus the overhead-valve engines supplied to SS-Jaguar.

THE DESIGN

By the end of 1957, Zebu design was well under way and was brimming with novelty: Dennis Mattocks was the project engineer, but Harry Webster, one of the great 'sketching' chief engineers of the classic era, had already briefed all his team, and guided them as to what he wanted. Not only had Webster decided to use a separate chassis frame layout, but his team had placed the transmission at the rear, Lancia Aurelia-fashion. The original style could not have looked more different from that of the Vanguard – not only was it a lot lower than the Standard, but it had more rounded lines, and a reverse-slope rear window.

In 1957 Standard was aiming to match anything that Ford could do, which explains this styling department study of an early Zebu mock-up being compared with a Consul/Zephyr Mark II. The existing Vanguard III is in the background.

By 1957 the small Standard four-cylinder engine – coded SC – was being built in huge numbers at Canley, and the six-cylinder that would eventually be used in the Triumph 2000 evolved from it.

Almost from the start, Standard-Triumph's stylists struggled to develop a cohesive shape for Zebu. This was merely one of the various front end shapes being tried at the time. In fact there were similarities to TR4 proposals of the period.

By the time Zebu came along, chief chassis engineer Harry Webster had started to develop independent rear suspension layouts. This was probably the first, tried on a TR3 sports car in 1957 and using a Zebu transaxle. In this case the half-elliptic springs were retained, with damper lever arms also acting as radius arms.

Right from the start, the Zebu chassis featured coil spring independent rear suspension, with semi-trailing links and a combined clutch / gearbox / final drive assembly mounted to the chassis frame. There was still a long way to go before the 2000 went on sale, but semi-trailing links for irs survived all that time, as did the chassis-mounted diff.

The very first style, produced by Standard without help from Michelotti, had pronounced peaks over the headlamps, vestigial fins at the rear, and that amazing rear window style. A comparison with contemporary Ford Zephyrs and the Vanguard showed wing crown lines several inches lower than on either car. However, it was thought to be all too extreme to carry forward into the metal, and by December 1957 several smoothed-out, alternative noses had already been tried.

In the meantime the chassis layout was being completed. Like the Herald, which had been designed a year earlier, the frame was of a backbone type with strong outriggers. The 80bhp six-cylinder 2-litre engine (complete with twin semi-down-draught Solex carburettors) was mounted well forward, but the combined clutch, gearbox and final drive assembly – nowadays we would call this a transaxle – was at the rear, the two being connected by a one-piece open propeller shaft. At that stage, an automatic transmission option was not even considered.

The transaxle incorporated a twin-plate clutch, the four-speed box gear casing was in cast alloy, while the rear axle casing was in cast iron. The engine flywheel, complete with starter ring, was actually mounted at the front of the engine, with the starter motor positioned to the right side of the cylinder block, driving forward to that ring. Once again, this was to eliminate bulk and bulges around the front footwells.

Other features included the use of TR3-type front suspension, matched to semi-trailing link independent rear suspension, where there were coil springs and lever-type dampers. Steering was by rack and pinion, there was a lengthy remote gear shift linkage, the handbrake was positioned outboard of the driver's seat, and inboard brake drums were mounted either side of the final drive unit.

When we started designing it [Harry Webster recalls], I insisted on better weight distribution than usual. So we put the gearbox and back axle together, which gave us much more space in the front instead of a great, big, hulking lump between the footwells. Mick Bunker did the design of the rear suspension. It was one of the deepest regrets of my life that we never patented this, because it was *the* first car with it. BMW followed us slightly later. [The original BMW 1500, in fact, was not previewed until 1961, though Fiat's rear-engined 600 of 1955 had used a cruder form of this suspension, so maybe a patent could never have been achieved.]

George Jones, who ran the transmission design team for many years, remembers that although the floor was indeed flat, the rear mounted transmission was under the rear seat cushion, which made the padding under the third, central, passenger (if carried) very thin. The layout itself caused many development heartaches:

> The whole concept created massive problems in relation to the engine/drive line. The prop shaft was always running at engine speed [this normally occurs only in direct fourth or fifth gear] and at first there was an inability to deal with the torsional vibrations which were set up, for you had the equivalent of two flywheels, one at each end of the prop shaft.

The tickover was very erratic at first, and to minimize it the prop shaft was given a big rubber joint at the front end and a Silentbloc joint at the rear. Although two-piece prop shafts were never considered, Jones investigated an early type of 'rope-drive' (an ultra-thin prop shaft, arced in side view, and supported in several annular bearings). In this, as in many other details, Standard-Triumph were pioneers, for this was years

By June 1958 Zebu's separate chassis design had reached this stage. There are superficial similarities to the Herald frame, but only in layout, for this was a much larger car. The six-cylinder engine had its flywheel and starter motor at the front, with the clutch mounted at the rear in the transaxle assembly; this was to ensure a flat floor to the passenger cabin. Note the TR3A-type of front suspension, though, with rack and pinion steering, the handbrake mounted outboard and the inboard rear brake location.

before the rope-drive layout was adopted by GM for the 1961-model Pontiac Tempest.

'We had torsional oscillations, at critical speeds,' Webster admits today. 'It was a bit of a job to solve, but we *did* solve it in the end.' Maurice Lovatt, who was much involved in 2000 development, remembers riding in the 'Zebu' prototypes and: 'I do recall vibrations. I remember sitting in the back and there was this objectionable vibration at cruising speeds.'

Even before the style settled down in mid-1958, Standard-Triumph knew that they were not going to be world pioneers. In the USA the 1957 Mercury Turnpike Cruiser had featured a reverse-slope window where the window could actually be retracted into the bodywork – but Standard still thought it could do a nicer job, overall.

Although the front of the car was still somewhat unco-ordinated, with a big scoop in the bonnet panel, semi-protruding head-lamps, and auxiliary lamps mounted above them, the side view was very smooth indeed; once you had accepted the reverse-slope rear

window, this was a modern, smooth-looking design, which would have outgunned the opposition from BMC, Ford and Vauxhall in the styling stakes.

Following the latest American fashions, the four-door saloon doors had frameless windows, in a hard-top coupé style. The monocoque layout on these cars featured a hidden half-pillar on each side, between the doors, the pillars being braced by a tubular structure across the back of the front seats.

DEVELOPMENT PROBLEMS

In the next two years Standard-Triumph struggled to make Zebu into a viable and attractive proposition, but they seemed to be fighting an impossible battle. Not only did they have to eliminate the drive-line vibration problems, but they still had great difficulty in settling an acceptable style.

As if that was not enough, there was also the enforced move of the entire design and development department (from Banner Lane to Fletchamstead North) to be faced, and the fact that the company ran into severe financial problems just when capital needed to be committed to tooling up for the new model.

In that time, five prototypes were constructed: X615 (1958), the original, firstly with a mocked-up frame, sent to Italy for the body shell to be constructed; X643 (1959), the second prototype; X646 (also 1959), an estate car version (I have never seen pictures of this car); X653 (1960), the fourth chassis, sent to Italy for Michelotti's revised body to be erected; X658 (1960), with a much modified chassis, and MacPherson strut independent front suspension.

By June 1958, Zebu had matured into this four-door pillarless saloon, complete with six-cylinder, 2-litre engine, and the reverse-slope rear window which (unbeknown to Standard-Triumph) was also being adopted by Ford of Britain. You might feel that the front end was still a little fussy, but all in all this is a very promising shape for the period.

Later Zebu prototypes had a revised four-headlamp nose, and there was ample space for the 2-litre, six-cylinder engine.

Michelotti had advised on the refining of the first style; it was not totally Michelotti, but there were many Italian touches. By mid-1959 it was all looking promising, though pre-occupation with the launch of the Herald – and the rather frantic search for a new TR style – meant that the timetable had already slipped considerably.

Tooling manufacture at Pressed Steel had not even begun, and there was no way that Zebu could have been on the market before 1961. The company was still committed to the reverse-slope rear window style – Harry Webster, Arthur Ballard (chief body engineer) and his team were all pleased with it ('we were tickled to death with it . . .') – so they invited the editor of *The Motor*, Christopher Jennings, to come along for a confidential preview.

Jennings, diplomatic as every magazine editor was in those civilized days, liked what he saw, but immediately found himself in a dilemma. His comments, the gist of which

have been related to me by several Triumph personalities of the period, were that: 'I think it's beautiful, but confidentially I must tell you that you'll be accused of copying, because there is another British car coming which will beat you into production by quite a long way, and it has got a cut-back window just like yours.'

The car in question was Ford's new Anglia Type 105E, and the larger Classic 109E which was to follow it up. 'It was a great shock,' Webster says, 'and because we knew we could rely on his word, we dropped the idea. So that was it. Start again, Webster!'

Three New Prototypes

This was when things got really frantic, and three different attempts – two serious and one only as a 'look-see' – were made to deal with the situation. One project was to reshape the rear of the existing style, while refining the front end and generally cleaning up the

---

Pressed Steel Co. Ltd

In the beginning, all car bodies were produced by coachbuilding methods, with metal panels fixed to a wooden skeleton. North American technology then made it possible for bodies to be welded up totally from pressed-steel panels. To bring this technology to Britain, Morris got together with Budd Manufacturing of the USA to set up the Pressed Steel Co., whose original factory was at Cowley, south east of Oxford. The first shells, for Morris Motors, were delivered in 1927.

Pressed Steel became financially independent in 1930, expanded mightily, and soon began supplying bodies (and, eventually, unit-construction monocoques) to most of Britain's 'big six' car makers. Standard took their first bodies from Pressed Steel in 1935 (the Flying Standards were not only built, but styled, by Pressed Steel), and became an ever-larger customer in the late 1930s. Triumph, the independent company, on the other hand, never built cars with all-steel bodies at this time.

After the war Standard-Triumph did no business with Pressed Steel for a decade, although they were by far the largest independent makers of car bodies. However, when the monocoque Standard Vanguard came to be designed in the early 1950s, the tooling and production contract was placed with Pressed Steel. Then, when the time came to replace the Vanguard by the Triumph 2000 in 1963, the planners once again chose Pressed Steel to build monocoques, a contract they retained throughout the car's career. Pressed Steel supplied complete saloon shells to Standard-Triumph and partly complete shells to Carbodies, for completion as estate car shells.

Before the Triumph 2000 came along, Pressed Steel had become a very large business indeed, with major factories at Cowley, Swindon, and Linwood (in Scotland), and in many ways it was almost integrated with the Morris (Cowley) plant. By the 1960s it was supplying body shells to BMC (Austin, Morris and derivatives of those cars), the Rootes Group, to Rover, Jaguar *and* Rolls-Royce.

Pressed Steel was absorbed by BMC in 1965, which temporarily made non-BMC customers nervous about future supplies, but by 1968, when they had become a part of British Leyland, most of those worries had automatically resolved themselves! Under British Leyland ownership, the Linwood factory was sold off (to Rootes/Chrysler), Swindon and Cowley were expanded, and Pressed Steel's identity disappeared into Pressed Steel Fisher, then into the maw of Leyland Cars.

By 1993 the original Pressed Steel factory at Cowley was the only surviving building in that geographical area, for the old Morris Motors complex had been demolished, and Rover cars were being assembled in what traditionalists and historians still called the 'Pressed Steel' factory.

---

detail, another was to produce a widened, lengthened Herald, and the third was to ask Michelotti to produce an entirely new shape.

The face-lifted (or perhaps I should call it tail-lifted) style of late 1959 saw a conventional rear window slope grafted on to the existing pillarless cabin, along with a smoother, four-headlamp nose, but although this was still a smart car it was no longer outstandingly different. Webster thought it was 'very difficult just to restyle the rear – somehow it would have unbalanced the whole car's lines, and in any case the tail was quite short . . .'.

The lash-up, which also dates from late 1959, was quite astonishing. In what has been described as a 'hot cross bun' manoeuvre, a Herald saloon shell was cut down the middle (lengthways) and across the top, widened, lengthened, and given four passenger doors, this mock-up then being compared with the Vanguard and a Ford Zephyr that the company had bought to study.

Unhappily, the result was predictable: it

Panic measures in the autumn of 1959! After the shock decision to ditch the reverse-slope rear-window style of Zebu, Standard-Triumph dabbled with the idea of matching a longer and wider Herald body to the existing chassis. Only a mock-up was ever constructed, but it shows how a completed car would have looked. Not thought distinguished enough, or different enough, it never got any further, though the four-door style, reduced in bulk, was later adapted for Standard Gazels (renamed from Heralds) to be built in India in the 1960s.

The board of directors of Standard in December 1959, just before financial problems began to batter their company. Marshal of the Royal Air Force Lord Tedder is at the head of the table, close to the camera, with managing director Alick Dick on his right.

still looked almost the same as the Herald. It was not a success. If ever there was a case of panic attack, this was it, and no prototype was ever built.

By April 1960, with Standard-Triumph sales slumping alarmingly, the directors had accepted that the Zebu project was in trouble but they persevered with it and allowed the projected launch date to slip to the autumn of 1962. The forecast for potential sales had been reduced to 25,000–30,000. Accurate estimates from Pressed Steel showed that it would cost £800,000 to tool up for Zebu body shell production, while Alforder Newton (a new company subsidiary) would require £131,000 to produce axles and disc brakes –

This was the finalized chassis for Zebu, different in almost every way from the 1958 variety, for it had become a pure backbone layout, and the gearbox had been moved forward to mate with the engine in the conventional manner.

The final panic redesign of Zebu came in 1960, when Michelotti was invited to style a new car. Completed in the late summer of 1960, this was the result; it was thought to be too bland, and was not liked. If you look carefully, you will see elements of Herald styling in many areas, notably in the front and rear wings, the roof, and in the (four-door) shapes of the doors from the Herald mock-up of 1959. Seasoned observers might also see slight resemblances to the BMW 1500 on which Michelotti was currently working, and which would emerge in 1961.

money which would be very hard to find. In the meantime the six-cylinder version of the Vanguard had been finalized, and would have to hold the line as an interim model.

A completely revised chassis layout was then produced – this one with an utterly different type of backbone chassis frame, with the gearbox now reverting to a conventional position behind the engine, and with MacPherson strut front suspension. This was X658, really the most practical of the prototypes, even though the flat-floor concept had been lost.

Finally, Michelotti was commissioned to produce an entirely fresh style, which need not retain any of the early lines. Completed in the summer/autumn of 1960, this was X653, and I have to say that by Michelotti's high standards it was a great disappointment, for it was angular where the original Zebu had been rounded, and brutal rather than delicate.

In some ways, X653 took the outline of the Herald as its inspiration, for some of the themes were repeated. There was a single, strong wing crown line from headlamps to

33

tail lights, the position and attitude of screen, door and rear window pillars were similar, and there was a square grille aperture in the nose. Alick Dick's team thought it was all too bland ('It wasn't a very good car, not a patch on the original'), and it was rejected. (Later, everyone realized that X653 was really an extension of Michelotti's current thinking on BMWs, because he had recently shaped the 600/700 types, and the first of the 1500s, both of which would be launched in the coming months.)

After that the end came suddenly, and decisively. At the board meeting of 19 September 1960, the Zebu project was cancelled, and at the same time the idea of using a version of its independent rear suspension under the TR4 project was also abandoned. Prototypes were cut up shortly afterwards – certainly when I first visited the experimental departments in the spring of 1961, no trace of Zebu remained – and the company was once again left without a successor to the Vanguard. The lumpy old Standard, for which sales had really dried up by the end of 1961, had to soldier on until mid-1963.

The Rambler Connection

The final alternative to Zebu appeared out of the blue in the autumn of 1960, and even on reflection this project seems to have been totally bizarre. It was so unlikely that collective amnesia seems to have afflicted some of those involved, Harry Webster included, who succinctly told me that he remembers nothing about it.

Even before the Zebu project was finally cancelled, Standard had been approached by American Motors of Detroit, who made an intriguing proposal. George Romney, the president of American Motors, offered Standard the use of the company's soon-to-be-revealed 1961 Rambler American shell, so that Standard could then transplant its own engines, transmissions and axles, and have a simple-to-develop alternative to Zebu!

According to AMC's then automotive boss, Roy D. Chapin Jr, the intention would have been for AMC to ship body components to Coventry in CKD (completely knocked down) form:

> I do recall Alick Dick and Martin Tustin, who came to Detroit in 1960, where they met

This was the 2-litre, six-cylinder engine, as finally unveiled in 1960 for the Vanguard Six, and therefore complete with semi-downdraught Solex carburettors. The Vanguard Six's gearbox (and, in this instance) overdrive would be further developed to suit the Triumph 2000 model.

After Zebu had been cancelled, the last pre-Leyland panic measure to replace the Vanguard Six was this hybrid – a 1961-model Rambler American saloon shell with which the Vanguard Six engine and transmission were mated. One car was built in the winter of 1960–1. In my opinion, it looked no better than the Vanguard it was meant to replace. No wonder it was abandoned.

with me and George Romney. Also I made a visit to Coventry, I believe, to view the assembly process on the Vanguard product – which was not particularly impressive. To my knowledge, while there doubtless were discussions, there were no conclusions on AMC distributing Triumphs in the USA.

It was bizarre because it was a totally unexpected corporate approach, and one which Standard had never considered up to that moment. For once there was not even much of an old boy connection to back it, for although Standard had links with American Motors in Australia, AMC's only previous manufacturing contact with a major British company had been when BMC (Austin) had assembled Nash Metropolitans in the 1950s, and even that deal had been set up by Nash before it became a part of the new American Motors Corporation.

Although AMC was much smaller than its neighbours in Detroit – GM, Ford, and Chrysler – in 1960, when the proposal was made, it was making no fewer than 422,000 Ramblers a year. The first Rambler American, a dumpy little six-cylinder engined car, had been launched in 1957, when it had a 100in (2,540mm) wheelbase and was no more than a warmed-over mid-1950s Nash, but for 1961 AMC had a

The six-cylinder engine was in volume production by 1962, this workshop study of a Vitesse 1600 unit confirming designer David Eley's philosophy of positioning every electrical component on the same side of the cylinder block.

	Vanguard	Rambler
	Six (1961 model)	*(1961 model)*
Wheelbase (in/mm)	102/2,591	100/2,540
Overall length, four-door (in/mm)	171.5/4,356	172.8/4,389
Overall width (in/mm)	67.5/1,714	70.0/1,778
Height (in/mm)	60.0/1,524	56.3/1,430
Widest wheel track (in/mm)	51.5/1,308	55/1,397
Engine (cc/bhp)	1,998/80 nett	3,205/90 gross
		3,205/125 gross
Transmission choice	4-speed manual	3-speed manual
(all manuals available	3-speed manual,	
with optional overdrive)	3-speed automatic	3-speed automatic
Unladen weight, four-door (lb/kg)	2,632/1,194	2,523/1,144

completely new model at the ready, which carried over little but the name.

This car, also set to run on a 100in (2,540mm) wheelbase, had an all-new unit-construction style shaped by Edmund Anderson (there were to be two-door and four-door saloons, coupés, a two-door convertible and an estate car) and in domestic form was powered by 3.2-litre engines of 90bhp (side-valve) or 125bhp (overhead valve).

Although the style itself broke no new ground, and had boxy proportions rather like the soon to be launched Rootes Super Minx/Vogue models, it was arguably at least as attractive and modern-looking as the best of the Zebus. American historian Richard Langworth later described the Ramblers as being: 'boxy and truncated, with odd, concave side sculpturing. Although they were genuine economy cars . . . they were anything but beautiful.' Further, in USA-made form, the American car weighed only about 2,523lb (1,144kg), which was actually 109lb (49kg) *less* than the Standard Vanguard Six, the interim model which was just about to be introduced.

Alick Dick, although startled by this proposal, was interested enough to mention it to his directors in September 1960, when he pointed out that such a car could replace the Vanguard Six at little or no capital cost. Since Standard had recently begun to lose oceans of money this was a very attractive thought.

Since both cars – Vanguard and Rambler – shared the same basic type of chassis, with coil spring independent front suspension, and a live rear axle, it is worth comparing the basic layout of the two models, to see how a 1961 AMC Rambler compared with the Vanguard Six (*see* table above).

The two cars, in fact, were broadly similar in bulk, weight and potential. If this project had gone ahead, Standard would have had little difficulty in transplanting its own new six-cylinder engine and transmissions into the new Rambler shell, without major body shell revisions being required. On the evidence of the speed with which Rootes had converted a Hillman Minx into a Singer-engined Singer Gazelle in 1956, this could certainly have been done within a year. Theoretically, therefore, Standard could have had a 'new' Vanguard in production before the end of 1961.

For a cash-strapped company, the Rambler American/Vanguard Six hybrid

The two Michelotti sketches that transformed Standard-Triumph's hopes for the future of its larger cars! Despite being two-door styles and titled 'New Herald', and dated May 1961, these were true ancestors of Barb, in every respect. Michelotti must have known that this shape was too ambitious for the smaller Triumph, but his themes were readily adapted to the larger Barb.

proposal was superficially attractive, but no one appears to have warmed to it after the first flurry of activity. During the winter of 1960–1 a single prototype was built up, a new-type American with Vanguard Six running gear, but no definitive styling proposals were ever worked up. Photographs (reproduced in these pages) prove the existence of the prototype, but no other trace remains.

Although Dick and Tustin actually flew to the USA in October to discuss this proposal further, nothing came of it, and as soon as Leyland made its take-over bid for Standard the project was speedily buried.

After Leyland secured Standard, and underpinned the future, it was time to think again. By the summer of 1961 a new project – the Barb – was under way.

3   The Birth of Barb

If it was the British government's credit squeeze of 1960 that helped to kill off the Zebu project, it was Leyland's take-over of Standard that inspired the birth of Barb. Although Barb – which became the Triumph 2000 – was a much better car than Zebu had ever been, it could not possibly have been put on sale without Standard's rescue by Leyland.

The fact is that by December 1960 Standard was staring bankruptcy in the face, for its finances had deteriorated rapidly throughout the year. In April the directors had been warned about falling sales – the Vanguard was dying fast, and the Herald's quality reputation was in the balance. By mid-summer sales had declined still further, for a world recession had joined that brought on in Britain by government policies.

In the last four months of 1960 Standard lost no less than £2.4 million, and its bank overdraft limit had to be extended. Alick Dick once said that 'money poured out like water. It was absolutely terrifying.' Then came the approach which saved the company, but cost many jobs: Leyland Motors offered to take over the company.

In what has been described as a 'whirlwind courtship', the Lancashire-based truck-building giant made the first serious contacts

---

Leyland Motors

The once-proud, Lancashire-based, truck-making concern began life in the 1890s as the Lancashire Steam Wagon Company, setting up shop in a village called Leyland, seven miles south of Preston. The Spurrier family, which still ran Leyland when Standard-Triumph was annexed, were in control right from the start.

Renamed Leyland Motors in 1907, the company expanded rapidly, even dabbling with the production of a luxury car (the Leyland Eight) between 1920 and 1923. Financial crises in the 1920s were shrugged off by 1928. Until the end of the 1930s Leyland concentrated on the building of commercial vehicles, often dabbling with the idea of taking over other commercial vehicle concerns, which included Daimler, Albion, Dennis and AEC.

After the Second World War, the company expanded rapidly, particularly in selling to export markets (Donald Stokes, ex-Lieutenant-Colonel in the British Army, was a star in this firmament, and joined the board of directors in 1953).

During the 1950s the company took over Albion (based in Glasgow) and Scammell (Watford), but still made no move into the private-car business. Talks with Rolls-Royce came to nothing, but chairman Henry Spurrier III (his grandfather had founded the business) was determined to broaden the base of his empire.

The first formal contacts between Leyland and Standard were made in the summer of 1960, on the basis of merging various overseas distribution businesses, but Spurrier's initial offer to buy Standard-Triumph and save it from bankruptcy did not follow until 14 November.

Leyland's formal offer to purchase Standard-Triumph was published in December 1960. There were no objections from shareholders, so financial details were rapidly agreed. The company formally took control in May 1961, when Spurrier took over the Chairmanship from Alick Dick.

in mid-October, chairman Sir Henry Spurrier made a formal offer in mid-November, and Alick Dick reported this to his board on 4 December. The world was

---

Stanley Markland

Rushed in by Leyland to sort out the mess they had inherited at Standard-Triumph, Stanley Markland rapidly turned the business round and saw it returning to profitability before being ousted by Donald Stokes in the Leyland power struggle of 1963.

Having joined Leyland as an apprentice in 1920, Stanley Markland quickly rose through the ranks. He was a capable production engineer, became Leyland's chief engineer in 1940, then became joint general manager and works director by 1951.

Ruthlessly logical, and known as an old-style authoritarian, he was already fully employed in September 1961 when Leyland drafted him in to Standard-Triumph as its new managing director – while retaining his other jobs, as managing director of Albion (of Scotland), and as works director of Leyland.

Under Markland, Standard-Triumph became a tight-spending concern, but it also began to invest in new models. Markland personally gave the go-ahead to the Spitfire project, the Vitesse, the Leyland 20 van, a start on the front-wheel-drive Triumph 1300, and was really the financial father of the 2000.

Under Markland, too, the company decided to get back into serious motorsport, first with the Spitfire, and also with the Triumph 2000.

When Leyland's chairman, Sir Henry Spurrier, fell mortally ill, Markland, as his deputy was already seen as heir apparent; but after he lost the chair to Sir William Black of AEC (which had just joined the group) he resigned at the end of 1963, leaving the motor industry for good and returning to his home in Lancashire.

---

informed of this approach on 7 December, when Standard agreed to merge with Leyland.

The corporate and political manoeuvrings which followed have been told many times before, and have no place here. Sir Henry Spurrier became Standard's chairman in May 1961, when Donald Stokes, Stanley Markland and Sydney Baybutt also joined the board. Before the end of the summer, though, Alick Dick had been ousted, almost all the original board members had also been dismissed, and Stanley Markland took over as Standard's managing director.

PREPARATION

In the meantime, with Zebu killed off and the Rambler-Vanguard idea discarded, Harry Webster had been told, informally, that he could start working up a new larger-car design, this time with a unit-construction shell. Straight away he approached Giovanni Michelotti, briefed him rapidly about a possible new programme, and set about producing a new model at great speed. Discarding the 'Z' list of four-letter names, Triumph decided to code this as Barb.

By the time Webster reported to the board in July 1961, his team had already schemed out the new model. He was confident that Barb could be ready for launch at the Earls Court Motor Show in October 1963. It was always understood that this car would be a Triumph, not a Standard, and at this stage it was to have been built with a 2-litre *and* a 1.6-litre engine.

Fortunately for Triumph, the projected programme was held, almost to the day, and the following is a summary of what was achieved :

April 1961   Michelotti
                    commissioned to style
                    the new Barb.

Giovanni Michelotti (1921–80)

Giovanni Michelotti was the Italian styling genius without whom Standard-Triumph's cars of the 1960s might never have come into existence. Starting with the Vignale Vanguard and ending with the Stag, his genius shaped a whole generation of Standards and Triumphs.

Born in Turin in 1921, he began his working life in 1937 with Stabilimenti Farina, the famous Turin-based coachbuilding concern. Not only was he a natural artist, but there was a genuine motoring tradition in the Michelotti family, for his father was a machining expert who worked for more than forty years in the industry.

Giovanni Michelotti was a sharp-featured, wiry, hyperactive and mercurial little man, with seemingly boundless energy and great ambition. People who knew him well say that he always worked very fast, and rarely refused a new commission. At times scraps of paper, menu cards, table-cloths and the backs of envelopes were all backdrops for his ideas.

He left Farina in 1949 to set up on his own as an independent stylist. He opened for business in a tenth floor studio in Turin, soon finding plenty to do for established styling

Giovanni Michelotti, the inventive Italian styling engineer who inspired so many fine Triumphs in the 1960s.

houses. In the first few years he worked for Vignale, Bertone, Allemano, Ghia and Balbo (often without his talents being publicly recognized by his clients). It was not until 1957, though, that a British businessman, Raymond Flower, introduced him to Harry Webster of Standard-Triumph.

Once he had demonstrated how rapidly and expertly he could turn ideas into completely detailed schemes, not only on paper, but as actual prototypes (his first accepted challenge was to offer a new shape on the Triumph TR3 chassis), Standard-Triumph put him on a long-term retainer (though not an exclusive one), which he held until the 1970s.

For Standard-Triumph, Michelotti styled cars like the Vignale Vanguard, the Herald/Vitesse family, the TR4 family, the Spitfire/GT6, the 1300/Toledo/1500/Dolomite range, the Stag – and of course the 2000 models.

Having refined the Vanguard Series III into the Vignale Vanguard he spent time working on Zebu models, but his elegant Barb shape was a fresh start, done quickly and sparingly, in the summer of 1961. All this, incidentally, was for a retainer of a mere £4,500 a year, plus payment of about £5,000 for each completed prototype!

In the same period he also shaped cars like the BMW 700, the BMW 1500/1800, the Daf 44/55, the Hino Contessa, the Triumph Italia coupé (which Vignale put into production), truck cabs for Leyland, secret work for several other makers, and a number of one-off styles for exhibition at motor shows.

After developing the 2000/2.5 Mk II shape, and that of the Stag, his influence over Triumph shapes fell away in the 1970s as the marque gradually disappeared into the anonymity of British Leyland and Leyland Cars. By the late 1970s his business was in trouble, when financial dishonesty on the part of a senior colleague was discovered.

He died in 1980, of a work-related disease, with his talents still not exhausted; he was only fifty-nine.

July 1961	Barb wooden mock-up completed.
November 1961	Style approval given, Pressed Steel Co. asked to make drawings and tool the body.
August 1962	All body drawings completed.
October 1962	First full-prototype body shell supplied (later shells provided at five/six-week intervals).
July 1963	First off-tools shell delivered by Pressed Steel Co.
August 1963	First pre-production cars completed.
October 1963	New Triumph 2000 introduced at the Earls Court Motor Show.

Speedy completion of body shell tooling was crucial to the entire project, and Pressed Steel (still an independent company at the time) could be proud of itself. At the time this was the tightest schedule that Pressed Steel had ever undertaken. PSCo, in fact, was anxious to secure new business from Standard-Triumph at this time, even though it had recently asked the board if the Vanguard could be withdrawn from production as the rate of building bodies was becoming uneconomic!

THE NEW CAR

The engineering team's first move was to sketch out a platform, engine/transmission positions, and suspension layouts. Arthur Ballard's body engineers then indicated the size of cabin that was required. Then, and only then, did Michelotti begin to shape the style that was to be so important to Triumph in the 1960s.

It was always understood that Barb would be longer, sleeker, lower and at the same time at least as spacious as the Vanguard that it was to replace, and here are simple comparisons of how the two cars differed:

	Barb (in/mm)	**Vanguard Six** (in/mm)
Wheelbase	106/2,692	102/2,591
Overall length	175/4,445	171.5/4,356
Overall width	65/1,651	67.5/1,714
Overall height	56/1,422	60/1,524
Cabin width (max)	54.5/1,384	54/1,372
Front seat headroom	36/9,144	36/9,144
Rear seat headroom	33/8,382	34/8,636

The whole of the cabin, in other words, was several inches closer to the ground, and although the new car was to be slimmer than before, there was to be no loss of elbow room inside.

Harry Webster, as always a great sketcher, soon established what would be in the new car. Except that there would be no separate chassis frame, much of the running gear would be an extension of what had evolved in the Zebu, some was an extension of what was shortly due to be launched in the TR4, and some was new for this car.

The Michelotti Style

The talented Italian stylist always seemed to work miracles when he was under great pressure. In no more than three months, not only did he personally shape the new car, but his team produced a full-sized wooden mock-up, complete with doors that could be opened, and delivered it to Canley for approval.

Although the new style looked good, it still took time to get it absolutely right. Maurice Lovatt, the experimental engineer who

Harry Webster

Nowadays, when computers, market research and number-crunching play such an important role in the design of new cars, there really is no place for personalities to develop. In the 1950s and 1960s, though, things were very different.

Once Standard's long-serving technical chief, Ted Grinham, had retired Harry Webster took his place, and within months the tempo and atmosphere of the engineering department had changed completely. From 1957, when he became Standard's Chief Engineer, until 1968 when Sir Donald Stokes persuaded him to move over to run the British Leyland design team at Longbridge, Harry Webster was the resident genius behind *every* new Triumph car.

Henry George Webster, always known as Harry, was born in 1917 and joined the Standard Motor Co. Ltd as an apprentice in 1932. During the Second World War he became deputy chief inspector, returned to the design offices afterwards and became chief chassis engineer in 1949. When Standard's technical director, Ted Grinham, became Alick Dick's deputy managing director in 1954, this left him with less time to look after engineering design, and after he retired (at the end of 1956), Harry Webster became chief engineer of the company, a position he was to hold with real distinction until 1968.

One of Harry's lasting claims to fame was that he 'discovered' Giovanni Michelotti in 1957, forged a great personal friendship with the Italian, and encouraged him to produce a series of elegant Triumph styles for the 1960s. Harry was not only an inventive designer but a capable administrator. He was also a true motoring enthusiast, dabbled in rallying from time to time, and was a fast driver who always liked to take a prototype home at night to see how it was progressing. He thought nothing of driving from Coventry to Turin and back during a weekend, just to see what Michelotti was cooking up.

During the Webster years Standard-Triumph struggled successfully to modernize their products, getting rid of dumpy bath-tubs like the Standard Vanguard and the Standard Ten, dabbling unsuccessfully with the Zebu projects before eventually settling on more advanced machinery. Not only were the elegant 2000 saloons and estates cars developed, and the 2.5PI derivatives added to that range, but the Heralds, Vitesses, Spitfires, GT6s and 'Michelotti' TRs all put into production. The 1300/1500/Toledo/Dolomite models also evolved, and before long the 2000-derived Stag Grand Tourer was also initiated. For Harry and his team this was an amazing record.

Although Sir Donald Stokes moved him to Longbridge in 1968, Harry stayed on as a director at Triumph until 1970, and he never lost interest in Triumph products. He moved out of British Leyland to become technical director of Automotive Products in Leamington Spa in 1973.

Finally, after retiring in the 1980s, he had more time to indulge Triumph enthusiasts and archivists who needed to top up their knowledge of the cars they loved so much. Even in the mid 1990s, when he still lived in the same Kenilworth house, he was as active and interested in all things Triumph as ever before. The TR Register did him the great honour of asking him officially to open its new Didcot HQ, where his speech proved that an old engineer does not have to be boring, and that enthusiasm for a marque need never die.

would look after most of the initial testing done on Barb, remembers that:

The one that went into production was very much like the one which came from Italy – but I do remember that the glass area was reduced. Harry Webster instigated that, and he also put in a reformed waistline at the bottom of the windows. The original mock-up had glass which came further down, particularly at the front screen. Someone told me that the glass area was reduced to save weight and cost. To me, though, it then lost a bit of the airy glassy feel, it took something off the Michelotti design.

John Lloyd (complete with megaphone, on the pit counter at the Le Mans 24-Hour race) was an unsung hero at Standard-Triumph. As the experimental shop manager, he led the development team that produced the first 2000s in 1962/1963. To the left of David Hobbs (dark sweater) is Lyndon Mills, sales manager at the time, while test-driver Fred Nicklin is further along the pit counter.

Harry Webster, outside the Vitesse convertible, and managing director Stanley Markland, were two of the most important contributors to the success of the 2000 in the early 1960s. It was Markland, a life-long Leyland man, who approved the massive investment in the new big Triumph.

The detail work and most of the structural engineering that accompanied it, was led by Arthur Ballard's team, with whom Vic Hammond and Leslie Moore concentrated on style work. Moore claims to have been responsible for adding the slightly overhanging roof panel at the rear, and the boot lid lip. Interestingly enough, pictures of the mock-up seen in the styling studio in October 1961 confirm that changes in clay had only recently been made in those areas.

As usual, styling approval had to wait until the directors had all seen it, paced round it, and given their opinions. They must have liked what they saw, for styling approval was granted in November 1961, after which Pressed Steel was immediately commissioned to produce body tooling. Even so, it would be many months before a complete body shell could be completed for testing to begin.

Body and Body Engineering

Although today's archivist sees the 2000 as a conventional monocoque car, he might forget that Standard-Triumph's engineers still only had limited knowledge of the technology involved. There had been two earlier monocoques – the Standard Eight/Ten/Pennant and the Standard Vanguard Series III – but much of the actual design work on those shells had been completed by Fisher & Ludlow and Pressed Steel respectively. Neither, of course, had been equipped with MacPherson strut front suspension, or independent rear suspension, so there was absolutely no experience in these areas.

Sturdy 'chassis legs' led back from the front of the shell to a massive cross-member under the front seats, and on the original cars structural members were also welded firmly to the floor pan to provide pick-up points for the semi-trailing arms: those would provide all sorts of grief in the development phase.

Massive structural sills linked front to rear wheel-arches, the overall result being a remarkably solid monocoque. Figures released when the 2000 was unveiled showed that the shell was one of the stiffest ever designed. Pressed Steel announced that torques of 6,000 to 6,500lb/ft (between front and rear axle lines) were needed to twist the shell through one degree. This made the shell two and a half times as rigid as the old Vanguard had been.

Mechanical Design

From 1957 to 1960 the ill-fated Zebu project had moved painfully towards maturity, only to be cancelled. Even so, the general principles of the late-model Zebu, complete with MacPherson strut front suspension, the semi-trailing rear, and the conventionally positioned gearbox features were all retained.

Mick Bunker of the chassis design office designed the semi-trailing arm rear suspension, which was considerably more sophisticated, in detail, than that of Zebu. Although the principle itself could probably never have been patented, the semi-trailing arms themselves were certainly unique at the time. To quote Webster:

> They were very cost effective. They were die castings [actually in LM6-M], and they incorporated so many things. You imagine if you wanted to do that with pressings, which is what would be done today? Getting one piece to do so many jobs was a great achievement.

'So many things' in one casting not only meant being a location arm, but included supporting the hub and bearings for the rear wheels, providing a housing for the rear springs, brackets for the hydraulic brake pipes and also offering a fixing for the telescopic dampers. It was neat, elegant, and throughout its life it was remarkably effective. Those who thought the castings could not absorb repeat shock loads (and there were plenty of shocks in this part of the car!) were confounded again and again.

Although a double-wishbone layout was considered, MacPherson struts were finally chosen for the front suspension, not least because these were ideal for spreading loads around the unit-construction monocoque. MacPherson struts, in any case, were becoming fashionable throughout the industry, which had realized that the layout was not only space efficient, but that the spring/damper loads could be cushioned by very strong panels close to the base of the windscreen.

Since the Herald and the TR4 already had rack and pinion steering, it was an obvious choice for Barb, but the use of rubber mountings was a surprise. Maurice Lovatt, who had worked on the Herald, commented that:

> We had some steering 'fight' on the Herald, which was quite objectionable on corrugations. We decided this was not acceptable for the man in the street, so mounted the rack on rubbers. Having done that on the Herald, it was accepted that we would do the same on the 2000. We're only talking about $^1/_8$ or $^3/_{16}$ of an inch of compliance, maximum. It also reduced impact loads throughout the steering gear.

Engine and Transmission

Except for the use of different carburettors and its installation, the engine was similar to

---

Six-cylinder engines ancient and modern

Although dewy-eyed traditionalists occasionally suggest that the Triumph 2000 engine of 1963 was related to older (independent) Triumph designs this was simply not true. Nor, for that matter, was there any connection with previous six-cylinder Standard engines.

The Triumph 'six' engines (1933–9)

When the Triumph Gloria was introduced in 1933, these cars used four-cylinder and six-cylinder versions of a Coventry-Climax engine design – 59 x 90mm, 1,476cc or 65 x 100mm, 1,991cc, which had overhead inlet but side exhaust valves. From the autumn of 1936 Triumph replaced this engine with its own new design, a simple overhead-valve design inspired by technical director Donald Healey, the six still being 65 x 100mm, 1,991cc. When the Triumph factory was bombed out of existence in 1940, this engine (and the tools to manufacture it) died with it. There was no connection, technical, visual or even philosophical, with the Triumph 2000 engine of the 1960s.

The Standard 'six' engines (1929–39)

Under Captain John Black's tutelage, Standard developed a closely related range of side-valve four-cylinder and six-cylinder engines throughout the 1930s; at first the benchmark dimension in all engines was a stroke of 101.6mm, this being increased to 106mm in the mid-1930s. Six-cylinder engine sizes ranged from 63.5 x 101.6mm, 1,930cc to 73mm x 106mm, 2,663cc. Such engines were also supplied to SS and SS-Jaguar.

Drawing on consultant Harry Weslake's experience, Standard also built overhead valve versions of the 'four' and 'six' for supply to SS-Jaguar, and used the overhead-valve 'four' in its own original post-war (Standard-) Triumph 1800 Roadsters and Saloons.

After the Standard Vanguard was introduced, this engine family finally died away in 1948. There was absolutely no connection with the Vanguard Six/Triumph 2000 engine which followed more than a decade later.

that used in the Vanguard, which meant that for the time being it was inflicted with a rather odd shape of cylinder head casting, along with porting and a combustion chamber that were rather resistant to power tuning. In the 2-litre Vanguard, 80bhp had been claimed but for Barb the published figure was 90bhp.

The first thing to note is that the six-cylinder engine was not mounted vertically, but was tilted over to the offside by 10 degrees. Naturally such a small inclination had no effect on the engine's height. Because of the battery position on the inside of the left-hand wheel-arch, the tilt was actually arranged to allow more servicing space around the engine auxiliaries. As designer David Eley reminded me about the original engine design: 'We wanted the distributor at rocker cover height for easy maintenance. It was also decreed that all major electrical components – starter, dynamo, coil, HT leads and plugs – were to be on one side of the

engine.' In later years, when Lucas alternators were fitted (these were bulkier, if shorter, than the dynamos they replaced), the designers must have been pleased about this.

Home-made Carburettors

Although the six-cylinder engine had only been in production for a year, the design team were delighted to get rid of the semi-downdraught Solex carburettors that featured on the Vanguard Six installation. 'We originally chose them to keep the engine height down,' designer David Eley told me. 'No one else ever chose to use them. I know why – they were a nightmare to us . . . they weren't the best carburettors in the world, I can tell you.'

Well before this carburettor fell out of favour at Canley, in fact, the company had taken the very bold step of designing its own carburettor. At a time when there were so many competent special designs available,

This six-cylinder engine bay shot (it is actually of a Vitesse 1600, but the bulk is the same) shows why it was necessary to angle the engine over, away from the camera, by 10 degrees to give clearance between the dynamo and what would be the battery tray in the Barb monocoque.

this was a very brave move. The result was a new type of constant vacuum instrument, with a diaphragm instead of a moving piston, which improved on anything SU had ever done: 'We felt this was the most logical way of doing carburation,' Harry Webster says,' and we weren't enamoured of the way SU did things. So we felt that the diaphragm way of doing it was superior – and we got round a lot of SU patents!' David Eley told me: 'The point was that the cost of SU carburettors to anybody who was not in BMC was astronomical. We did appreciate the benefits of an SU-type of carburettor.'

Soon after Standard-Triumph absorbed Alford & Alder of Hemel Hempstead, technical director John Lind was consulted. There was a bright young man there called Dennis Barbet, who was charged with the job of designing a carburettor. An engine test cell was set up at Alford & Alder, complete with Heenan & Froude water brake, and in a remarkably short space of time promising prototypes were in existence.

By 1960 Standard concluded that it was not equipped to make such a carburettor in its own factory, so set about finding a partner. SU, naturally, were not approached at all, and it was not until Amal (the Birmingham-based motorcycle carburettor company, owned by IMI) had turned it down, that Zenith-Solex became involved.

Although Barbet had already produced an ingenious new design, which skirted every SU patent while retaining the variable choke system, a lot of detail work remained to be done. Zenith went back to their parent company, Bendix-Stromberg of the USA, Triumph TRs were sent out there, with SUs and prototype Strombergs to swap from engine to engine (which explains why the original CDs always fitted the same manifolds as SUs did!), and in a matter of months the right results were available.

The company therefore handed over the design to Zenith-Solex in Europe on the basis that Standard-Triumph should then have exclusive use of the new design for a period, which I believe to have been twelve months. Details of any royalty agreements for other supplies have never been published.

Transmission
At first glance, all traces of Zebu design had been eliminated from the new car's transmission, though designer George Jones reminded me that a modified version of the Zebu back axle was used under Barb. Jones, originally an Austin-Longbridge trained engineer, had worked at Standard since 1948, originally under technical boffin Lewis Dawtrey, and knew more about transmission design than almost anyone in the industry. He remembers that capital for new tooling and brand-new designs was always scarce at Standard, which explains why there were still traces of the Vanguard gearbox in that chosen for Barb:

> We started with three speeds and a column change, then went to four speeds with a floor change for the TR2. Then we had three *or* four on the floor for the Vanguard Six. Basically it was always the same concept. We kept on using what tooling we could.

When adding synchromesh to first gear for the TR4 (previous TRs, and the Vanguard Six had used a 'crash' first gear), the length of the casing had increased by a mere 0.44in (11mm), but most of the gear wheels, synchro cones and bearings were unchanged.

As on the Vanguard, Laycock de Normanville overdrive was optional (it operated on top and third gears), as was Borg Warner Type 35 automatic transmission.

PROTOTYPE DEVELOPMENT

Even though Barb had the highest possible priority in all departments, it took ages to get

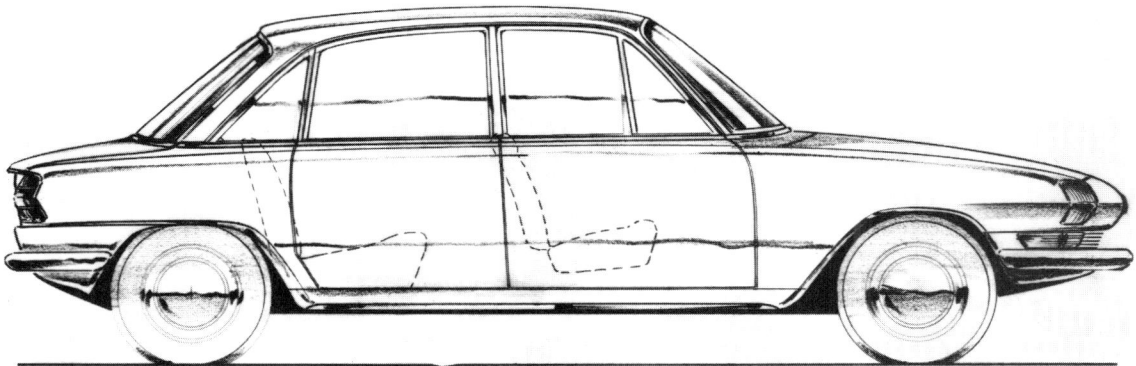

One of several sketches produced by Michelotti in 1961 to show how he saw the shape of the new Barb. According to Harry Webster, the Italian would probably have produced this sketch in less than half an hour! Except for the truncated tail, this already approximates to what was chosen.

By the autumn of 1961, the Michelotti Studio had built this full-size wooden Barb mock-up for refinement in Coventry. Compared with the production car, the only significant changes still to be made were in the profiling of the headlamp recesses. Triumph's Leslie Moore had already added the peak over the rear window (that section of the mock-up is smeared with clay). Note the prototype Vitesse bonnet on the left.

49

Well before Pressed Steel could start to deliver complete body shells, a prototype Barb platform, scuttle and wheel boxes was built, to which all the running gear was added. Triumph's 'can-do' craftsmen then added a cheap and cheerful bird cage structure to ensure rigidity, covered the engine bay and the passenger cab with soft-top roofing material – and testing could begin in the summer of 1962. This was X710, or the 'birdcage' as it became known. Look carefully and you will see the Triumph Herald steering wheel, instruments and seats. That is a proper prototype 2000 fuel tank between the rear wheelarches, and there are hefty bags of sand in the rear footwells to bring the assembly up to full weight.

The very first Barb/2000 body underframe of 1962, as seen on its side in the experimental department at Fletchamstead North, before it was built up into X710, the 'birdcage'. The most significant feature here is that there was no provision for a separate rear cross-beam. Instead, the angled mountings for the semi-trailing rear wishbones were welded into the underframe, tying up at the outside to the sills and near the centre of the car to 'chassis legs' surrounding the rear differential position.

the first complete prototype on the road, as many of the panels would have to be supplied by Pressed Steel. Maurice Lovatt and Dennis Mattocks had to wait, impatiently, before 'half a prototype' (as someone once described it) took to the road.

Also succinctly nicknamed the 'birdcage' this used the pressed-steel underpan/platform of the new car, to which a rudimentary tubular-steel structure was welded up, braced to give some semblance of rigidity. There were inner wings but no outer wings, the 'bonnet' and 'roof' were in plastic soft-top material fixed to the spidery structure by pop-fasteners. This unappealing mixture was topped off by the use of Herald front seats, Herald instruments and steering wheel, a fuel tank mounted between the rear arches, and a heap of sandbags in the rear footwells to bring it all up to estimated full-load weight.

This remarkable machine, which was very important in Triumph 2000 history, but would have won no prizes for looks or even character, carried the experimental commission number of X710. Although it carried front and rear lights, and a space for number plates, it was never road-registered (even in those days I cannot see any licensing authority deeming it road legal!), and completed most of its work on private ground, especially at the MIRA proving grounds.

> I took this thing to MIRA [Maurice Lovatt told me] with Harry Colley. I took it on the pavé, along with a Phase III Vanguard for comparison, to do an assessment of the ride. The Phase III was skittish, all over the place. Then we went down with that – the underframe – and it was just unbelievably better. Compared with the Phase III Vanguard, it was transformed.

That was in the autumn of 1962, but the first completed car (X711 – still with very bare interior equipment) was not ready for several weeks, which meant that road testing of this very important machine did not really begin until nine months before the first production cars were due to be assembled.

At first X711 was fitted with a 1.6-litre Vitesse engine, which is a reminder that Leyland had originally asked for 1.6-litre and 2-litre versions of the car. Not that Leyland, still very new in the car business, always seemed to know what was practical – soon after they had taken control at Canley, they asked for a Herald 1200 engine to be installed in a Vanguard for assessment. It was done, and the results were embarrassing.

Leyland was constantly pushing for costs to be lowered, and in November 1961 Donald Stokes asked for the 1.6-litre to sell for £850, and for the 2-litre car to sell for £900. Standard's cost experts bit their collective lip, and the designers soldiered on, but these seemed like ludicrous figures for what was already promising to be a technically advanced new model.

Even though Pressed Steel was persuaded to reduce its body shell prices from £105 to £94, no one ever thought Barb could be sold for less than £1,000; the Vanguard Six, by the way, cost £1,051 in 1961–2.

Although some work was done on the 'Triumph 1600' – which would have been distinguished from the 2000, visually, by using a different front grille, wheel trims and decoration – not even the sales force took the idea seriously. Not only would it have been sluggish and under-powered (65bhp, perhaps, compared with 90bhp for the 2000), but it would probably have used even more petrol if owners had tried to urge it along.

Long before the 2000 was launched, the 1600 had been dropped, and X711 had been re-engined, to the relief of all concerned. Donald Stokes must have accepted other opinions, too, because little resistance seems to have been given to its cancellation.

Good News and Bad News

During the winter and spring of 1963, Pressed Steel delivered five prototype saloon body shells, but just as testing began in earnest the British weather clamped. January to March 1963 was the period of the worst winter in living memory, which meant that endurance driving was badly hit and, worse, detail assessment was delayed again and again.

The good news, however, was that Barb was proving to be an exceptionally strong car on rough roads, especially on awful Belgian pavé surfaces where 1,000 trouble-free miles (1,600km) was considered an excellent pass-off target. By May 1963 a prototype had completed nearly 2,000 miles (3,200km), which was not only a record for Standard-Triumph, but was believed to be a record at MIRA at the time. The only problems were cracking close to the bonnet hinges on the front cross panel.

The bad news was that prototypes were not nearly as refined as they needed to be for the new car to be a success. As with Zebu in the late 1950s, there was a quite unacceptable amount of vibration and resonance set up in the shell, and with only months to go before Pressed Steel was due to begin supplying bodies in big numbers, an air of panic descended on Fletchamstead North.

When analysing the design in its March 1964 issue, *The Automobile Engineer*'s experts summed up thus: 'In the original design, the members that carry the [forward] pivots of the suspension arms were welded directly to the floor of the body structure, but when prototype testing began it soon became apparent that some sort of sub-frame would be required to reduce road noise . . .' Which is precisely how Maurice Lovatt remembers the problem: 'I recall standing around Ken Rose's drawing board, with a Pressed Steel man alongside, seeing how we could modify the shell. I remember contributing by saying

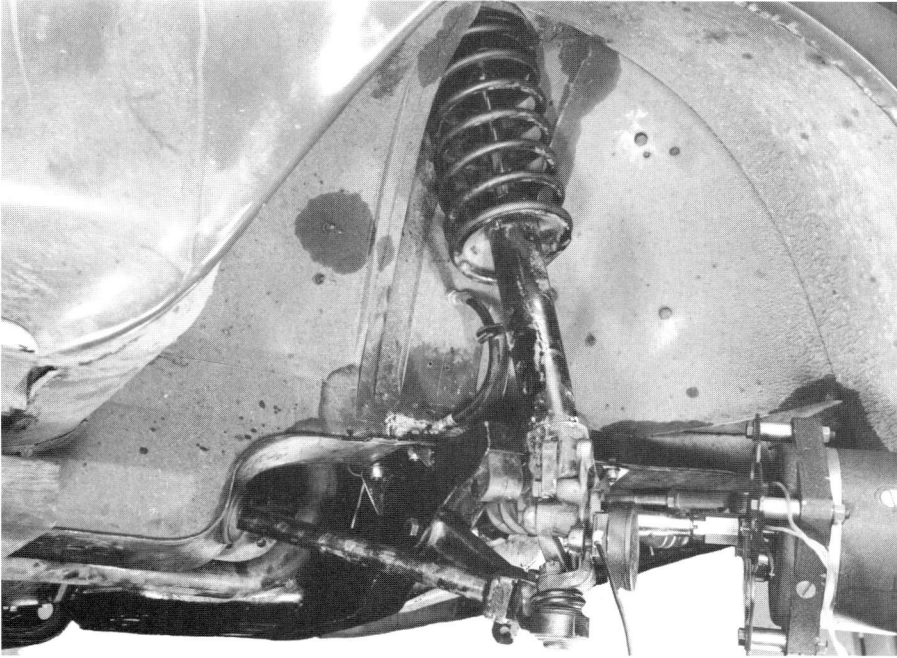

Rig testing of the Barb/2000 strut front suspension in progress. This compact MacPherson strut layout proved to be robust, even in ultra-rough rallying conditions

that a cross-beam would be advantageous from the noise point of view.'

Lovatt and Mattocks had already tried rubber bushes in several different places, but in the end Harry Webster had to authorize major changes to the underside, which Pressed Steel somehow managed to incorporate without delaying the schedules.

Instead of using welded-on members (as seen in the picture of the very first prototype floor pan on page 51), a V-shaped steel cross-beam was used instead. By mounting this to the shell on rubber bushes, and additionally by spreading the twisting load of the rear axle casing, also mounting that on rubber, the problem was solved. Instead of being totally unacceptable, it was smooth and very definitely saleable; Triumph later found out that they had solved a problem encountered on Rover's new P6, but that Rover had not produced as elegant or as effective a solution.

Even so, to bring a major new model into the market place so quickly, and with so few

prototypes, was astonishingly brave. To achieve all this, and to ensure that the early cars were also reliable, refined and well finished, was a remarkable achievement. At the time, probably no other tightly knit team of engineers, production specialists and planners could have done so.

Much of the detail work in 1963 went into making the car as smooth, refined, and trouble-free as possible. Before he moved on to another job, Maurice Lovatt recalls doing much work on ride and handling matters, proving to everyone's satisfaction that a soft ride and good roadholding could still be achieved at the same time. Later, after the Triumph *and* Rover 2000s had been put on sale, Triumph was gratified to find that its own installation was the better of the two. 'One of our problems,' Lovatt says, 'was that in the early days there was a quality problem with the wheels. They didn't show as out of balance, but the tolerances weren't tight enough, we were getting ovality problems.'

List of all Barb/2000 prototypes built from 1962 to 1965

Commission Number	Details
X710	Underframe and running gear only, with tubular-steel superstructure. Never converted to complete-car status.
X711	First prototype, only prototype with 1.6-litre engine. Registered 5264 VC in 1962.
X713	Second prototype. Registered 9081 VC in 1963.
X716	Third prototype. Registered 9082 VC in 1963.
X720	Fourth prototype. Registered 5384 KV in July 1963.
X721	Fifth prototype. Registered 2418 KV in July 1963.
X723	Styling studio car, not shown as ever registered, actually a car built from off-tools parts by Production Engineering, and handed over in January 1964.
X729	Estate car prototype. Registered 698 CWK in May 1964.
X740	Estate car prototype. Registration number not recorded, but put on the road in April 1965.

The finished product. In fact this is the third prototype – X716 – as used for publicity record shots in the summer of 1963.

The third prototype, X716, had been thoroughly updated, refurbished, and made to look as new when this study was made in the summer of 1963. This was just the start of a long-running success story for the 2000 range.

The finalized 2000 interior/facia, of X716, the third prototype car 9082VC, in its ready for production state. Gradually, over the years, the instrument display would get more complex and the panel rather more cluttered. Compared with the old Vanguard, though, this was a miraculously svelte layout.

Once complete prototype and early-production cars were available, concentrated testing was carried out at the MIRA proving grounds. Gordon Birtwistle lost count of the hours he spent at MIRA on the handling circuit and on the various special surfaces.

Surprisingly enough there were few handling problems to be overcome. Considering that cross-ply tyres were to be standard, and the amount of what engineers call 'swing axle effect' was endemic in the semi-trailing rear suspension layout, this was a great relief.

There was, in fact, considerable camber change in this layout – from 5 degrees positive on full rebound to 7 degrees negative on full bump. Great care was needed to keep the length of the coil springs inside design limits. Nominally, the rear wheels took up a 1-degree positive camber when unladen at the kerbside, but when the car was fully loaded this had changed to 3 degrees negative.

Strangely enough, the development team seemed to have few problems with 'stiction' in the splines of the rear suspension drive shafts; this would occur, with irritating consistency, after the cars had gone on sale.

Then there was Leyland's newly imposed engine flexibility standard. Stanley Mark-

land, the vastly experienced truck engineer, was not at all impressed by the Solex-carbed Vanguard Six engines he had tried. Engineering's elder statesman, Lewis Dawtrey, was charged with the development of Zenith-Stromberg carburation, so that it would always be impeccable.

One of Markland's requirements, proved on a steady uphill slope on a public road near Coventry, was that the car should pull away on full throttle in top gear from 10mph (15km/h)! He would not accept any fuss, hesitation, bother or spitting – and within reason (for this was a six-cylinder engine, after all) this was finally achieved. This was, however, an entirely new culture to Standard-Triumph engineers.

Trim and Furnishing

Harry Webster always used to complain that the only part of his job he did not enjoy was setting budgets. He could also have added that any number of his team's design

Leslie Moore's part in developing the style, and particularly the interior style, of the 2000 should never be underestimated. Moore, here seen shaping a prototype sports car model, shunned publicity throughout his working life and was happy to see Michelotti receive most attention.

Triumph 2000 (Mk I) (1963–9)

Produced
August 1963 to September 1969

Identification
Chassis numbers carried the prefix MB

Layout
Unit-construction body/chassis structure in steel. Five-seater, front engine/rear drive, sold as four-door saloon or five-door estate car

Engine
Type	Standard-Triumph six-cylinder
Block material	Cast iron
Head material	Cast iron
Cylinders	6 in line
Cooling	Water
Bore and stroke	74.7 x 76.0mm
Capacity	1,998cc
Main bearings	4
Valves	2 per cylinder, pushrod and rocker operation
Compression ratio	8.5:1
Carburettors	2 Zenith-Stromberg 150CD
Max. power (net)	90bhp @ 5,000rpm
Max. torque	117lb/ft. @ 2,900rpm

Transmission (Manual)
Clutch	Single dry plate, 8.5in diameter, hydraulically operated

Internal gearbox ratios
Top 1.00, 3rd 1.386, 2nd 2.100, 1st 3.28, reverse 3.369
Final drive   4.10:1
16.9mph/1,000rpm in direct top gear
Optional Laycock overdrive (on top and third gears) had a ratio of 0.82:1, overall top gear ratio of 3.36:1
20.6mph/1,000rpm in overdrive top gear
Note : Final drive ratio of 3.7:1 became optional from spring 1966

Automatic transmission (optional)
Torque converter	Maximum torque multiplication 2.0:1

Internal Transmission ratios
Top 1.00, intermediate 1.45, low 2.39, reverse 2.09
Final drive   3.7:1
18.75mph/1,000rpm in direct top range

Suspension and steering
Front	Independent by coil springs, MacPherson struts, lower wishbones, telescopic dampers in struts
Rear	Independent by coil springs, semi-trailing wishbones, telescopic dampers

Steering	Rack and pinion
Tyres	6.50x13in cross-ply (saloon)
	175x13in radial-ply (optional on saloon, standard on estate)
Wheels	Pressed steel disc, four-stud fixing
Rim width	4.5in

Brakes
Type	Disc brakes at front, drum brakes at rear, with vacuum servo assistance
Size	9.75in diameter front discs; 9 x 1.75in wide rear drums

Dimensions (in/mm)
Track	
Front	52/1,321
Rear	50.4/1,280
Wheelbase	106/2,692
Overall length	173.75/4,413
Overall width	65/1,651
Overall height	56/1,422
Unladen weight	(saloon) 2,576lb/1,168kg
	(estate) 2,688lb/1,219kg

Note Between 1965 and 1968 AMI (Australian Motor Industries) of Melbourne, which assembled 2000s from CKD packs, also sold a special version of the car badged 2000 MD, where MD stood for managing director. Compared with the standard car, this had overdrive as standard, and chrome-plated centre-lock wire spoke wheels; Dunlop SP radial-ply tyres were standard equipment. Wheels came from an Australian suppler, and increased the wheel tracks by 2in (50mm) two inches.

A wood-rimmed steering wheel was fitted, along with extra VDO instrument pads above the normal instrument panel, which contained an electric clock and a rev counter. A vinyl roof covering was also available, and on a very few late-model cars a triple Stromberg carburettor engine was also offered.

The 2000MD was no more than an 'Australian special', made by converting partly complete cars from the Melbourne assembly tracks. Originally it was planned to build 400 cars but demand was low, and only about fifty units were ever sold. AMI has now been out of business for some years, and no further information is available.

ambitions were frustrated by a limit on what could be spent.

When Arthur Ballard's team came to furnish the 2000, they faced a near-impossible task. Whereas they wanted to make this the most luxuriously trimmed car that Standard-Triumph had yet produced, there were serious cost constraints imposed by Donald Stokes's sales people.

In the end, the interior looked plushy and well equipped, with a modicum of good-looking carpet, wood veneer, and large seats. The front seats reclined, there was a fold-up vanity mirror inside the glove box lid, and indicator repeaters were fitted to the door pillars between front and rear doors. Even so, some compromises had to be made. Plastic seat covering had to take the place of leather (which was an optional extra), a clock could not be afforded, and for the time being the heating and ventilation system was still quite simple.

Leslie Moore's team shaped the facia of the original Triumph 2000. This was a 1962 proposal, close to what was eventually chosen, except that the instruments are very different and there are face-level vents on each side of the ashtray, which were not adopted.

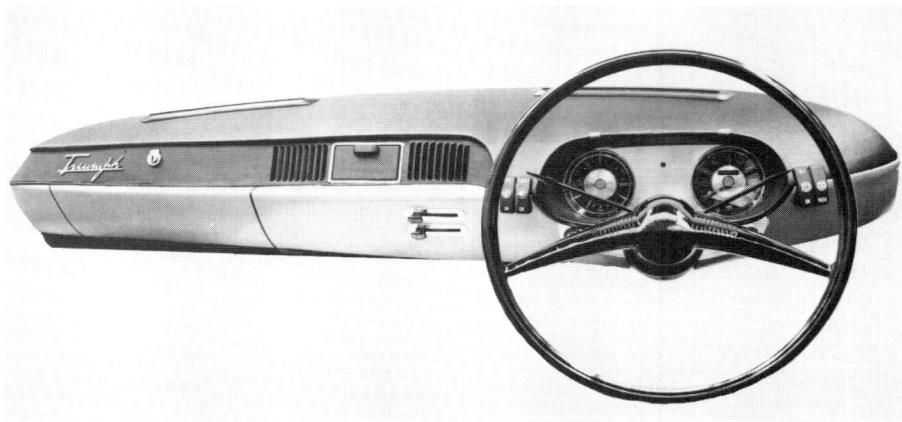

BUILDING THE CARS – A MECHANICAL JIGSAW

By mid-summer 1963 Barb had officially become the Triumph 2000, all the pre-launch publicity pictures had been taken, and it was almost time to 'go public'.

In May Pressed Steel had promised that it was cutting the holiday shut-down for the new section of its factory in Swindon, and that it would deliver the first handful of body shells by early August, and that 400 shells would be delivered by the end of September.

It was enough for Stanley Markland to authorize a production rate of 20,000 cars year – about 400 cars a week. In the meantime, space had been cleared at Canley for 2000 assembly to begin. The last of the old-type Standards – an Ensign De Luxe – had rolled off in May 1963, and everyone was itching to get their hands on the new cars.

Naturally, final assembly would have to begin around a complete unit-construction body shell. The Pressed Steel Co. started by

As soon as the car went into production Michelotti took delivery of a 2000. Naturally he began modifying it straight away, one result being the fitment of centre-lock wire spoke wheels, and special badging on the rear quarters.

Canley – the new assembly hall

As Standard's Canley complex was developed, final assembly of the cars was always concentrated on the south side of the site, in a hall close to the Coventry–Birmingham railway and Canley railway station. It was not until the end of the 1950s that a modern, more spacious, assembly hall was planned.

First discussed at board level in April 1958, the new hall was built using the proceeds of the sale of the Ferguson tractor assembly business; at the time it was estimated to cost £2 million. The new assembly hall rose just 200 yards further west of the older buildings, on previously undeveloped ground. Intended to take in painted shells and to turn out completed motor cars, its layout featured several parallel assemblylines, and when Standard-Triumph prolixity was at its height it was possible to see Heralds, 2000s, TRs, Spitfires and 1300s all taking shape close to each other.

Construction of the new facility began in 1959, and the building was originally meant to be ready by October 1960, but Standard's financial crisis delayed completion. Car assembly began in the new hall in March 1961, and by 1962 the old facility had been cleared and used for other purposes.

Costing £2.5 million when new, the new hall had every possible modern control and sorting detail (which included TV surveillance of body shell sorting), and before final assembly began work continued on several levels. When the 2000 model was introduced, the new building was already busy with the erection of Heralds, Vitesses, TR4s and Spitfires, though the last Standard-badged cars (Ensigns and Vanguard IIIs) had disappeared months earlier.

Except for CKD (completely knocked down) kits sent abroad for assembly overseas, every car in the 2000 family was built in this building, though engines, transmissions and other components all flowed in from a variety of sites. The last 2000 models were built there in May 1977.

The last Canley-built cars of all to take shape in this vast hall were Spitfires, TR7s and Dolomites. Assembly closed down in the autumn of 1980, after which the building was put to more mundane uses by Austin-Rover. In later years it was given over to Unipart, who expanded their parts supply operation into it.

The vast assembly hall at Canley, newly built in 1959–60 and fully open by 1961, was where the 2000/2.5 family was built throughout its life. This shot, featuring Spitfires and TR4s in the mid- and background, pre-dated 2000 assembly by a few months. By the 1980s all car assembly in this monolith had ceased.

This well-known aerial shot shows the entire Canley/Fletchamstead complex, as it was when the 2000 was introduced in 1963. This shot faces east. Coventry city centre is out of shot at the top of the picture, and the Coventry by-pass (A45) crosses the base of the picture. The assembly hall is to the right centre of this shot, with the ageing administrative headquarters (and 'Ivy Cottage') beyond it. Fletchamstead North, the design/engineering/development block, is at the bottom left, and vast machine shops, fronted by the sales block, are to the bottom right. It all looks very different today!

pressing all the panels, then welding them together, at a new factory in Swindon, before delivering them as bare shells by the transporter load – six at a time – to Triumph's ex-Fisher & Ludlow factory at Tile Hill, a western suburb of Coventry.

There, along with every other Triumph body shell, this thoroughly modern plant would subject the monocoques to an eight-stage treatment, first protection with phosphates, then alkyd dip priming, and finally priming, undercoating and painting. The gleaming new shells would then be loaded back on to the purpose-built transporters, driven due east for two miles to Canley, there to be rolled in to the southern end of the vast modern assembly hall close to the Coventry–Birmingham railway line.

In the meantime, major components would arrive from many different directions –

George Turnbull (1924 – 1992)

A Coventry-born boy who started his working life as an apprenticed Automobile Engineer at the Standard Motor Co. in 1941, George Turnbull ended his days as a knight, and as chairman of the Inchcape Group. Along the way he had worked in the top management of Standard-Triumph, British Leyland, and several other industrial concerns.

Except for a short spell with Petters in the 1950s (a diesel engine manufacturer), George Turnbull worked continuously with Standard-Triumph until 1968, when both he and Harry Webster were invited to move over to Longbridge to help sort out the Austin-Morris side of British Leyland.

At Standard, he became technical director Ted Grinham's personal assistant in 1947, ran the experimental department for a time in 1954–5, then (after his sojourn at Petters) became production manager of the entire Canley complex. By 1961 he was on the board of Standard, staying there as director/general manager under Stanley Markland and Donald Stokes until 1968. During that time he had a very happy and fruitful relationship with Harry Webster, the two of them effectively being Standard-Triumph's own product planning experts throughout the period.

It was under Turnbull that the Barb/2000 project was finalized, put into production, and put on sale, and before he moved across to Longbridge as managing director of Austin-Morris, he also saw the restyled Mk II project get under way.

After leaving Coventry, Turnbull stayed at Longbridge for five years. Expecting to take over from Lord Stokes as chairman when the Leyland man retired, he walked out when he saw John Barber winning that power struggle. In 1974 he moved to Korea to set up the Hyundai car-making business, and three years later he spent time with Iran National, which assembled Hillman Hunter cars. From 1979 he became chairman of Talbot UK (later Peugeot-Talbot UK), then in 1984 he joined Inchcape, where he became chairman in 1986.

Knighted in 1990, he retired from Inchcape in 1991 but enjoyed only a short rest. When he died in 1992, aged sixty-eight, he had an almost unmatchable record of top management in the world's motor industry.

engines from another building at Canley, gearboxes from elsewhere on the Canley site, and axles from the Radford factory (two miles away, hard by the Daimler plant), while Laycock overdrives would arrive from Sheffield, and Borg Warner automatic transmissions came from Letchworth. Front and rear suspension components would arrive from Alford & Alder in Hemel Hempstead, with suspension struts and dampers coming from Armstrong Patents in Yorkshire.

The roads from Birmingham to Canley were going to be busy, for electrical equipment would come from Lucas, glass from Triplex, body fittings from Wilmot Breeden and tyres from Dunlop, while castings for cylinder blocks, cylinder heads, gearbox and axle casings all came from Beans Industries, also in the Birmingham/Wolverhampton area. For the time being Beans would also assemble complete engines, though this process would be transferred to Canley in future years. Lockheed/Borg & Beck of Leamington Spa provided brakes and clutches.

This was, in effect, the multi-million pound manufacturing challenge at which Standard-Triumph was so adept. When the lines began to move in August 1963, slowly at first, then with ever-increasing urgency, Standard-Triumph and Leyland could only wait and hope. Was it what their public would want?

4 The 2000 on Sale: 1963-9

The new car was finally introduced in October 1963, when Barb at last got its proper name – Triumph 2000. The press gave it a friendly, even enthusiastic, reception, but at the time none of the pundits realized just what sort of a miracle had been achieved to meet this date. In little more than two years a good idea – a wooden mock-up, no more – had been turned into a production car.

There are still cynics who suggest that the launch was premature, and that in October 1963 Standard-Triumph was not yet ready to start delivering cars in big numbers. That may be so, but the fact is that the first off-tools motor car (today, no doubt, we would describe it as a pre-production machine) was finished on 12 August 1963, and that more than 1,500 production cars were actually built before the end of calendar year 1963.

Standard-Triumph was always under pressure to get the new car into production, not least from the dealers, for the last Standard in the Vanguard/Ensign range had been produced three months earlier, at the end of May. Even so, even though all the facts are well known today, it is still remarkable to note how rapidly this major project had come to fruition. As already noted, Pressed Steel had first been consulted about body tooling in November 1961, yet it had delivered the first off-tools bodies twenty-one months later. By 1990s standards that would be difficult, but in the early 1960s it was nothing short of miraculous.

ROVER AND TRIUMPH – RIVAL 2000S

Much was made of the fact that both Triumph and Rover had introduced a new up-market 2-litre car at the same time. What was not spelt out at the time was that each company had known what its rival was up to for several years, and both were aiming for the same type of customer. When top managers had briefly co-operated during merger talks in 1959, their forward programmes had been discussed, though little detail was provided by either side. Since industrial espionage was little practised at the time, rival engineers could only learn by what they saw of prototypes at the MIRA proving grounds, near Nuneaton.

Rover, in fact, had started work on its P6 project at least three years before work on the Barb even began. In many ways, the two cars were technically very different, and because of this there was certainly space for both of them in the market place.

Since Spen King was a top engineer closely involved with both cars – in at the birth of the Rover 2000, and in the continuing development of the Triumph 2000 after 1968 – it is fascinating to hear how he compared the two:

> I think both of them had their failings. I think the 2-litre six-cylinder Triumph engine was beautifully smooth, more refined than the Rover four-cylinder engine; I do think that the Triumph engine was very superior.

The rival 2000s – Triumph versus Rover

For years before the Triumph 2000 was announced, Standard knew that Rover were developing a potential rival to it, for the two companies had discussed future programmes during merger talks in1959.

Rover, in fact, took much longer to take their 2000 (P6) project from concept to series production than Triumph would have deemed acceptable. The very first P6 was sketched in 1956, the first full-size clay models followed in 1958, and prototypes were on the road before 1960. An entirely new factory block was completed in 1962, and the P6 assembly lines began to move, albeit very slowly, in the spring and summer of 1963.

Although the two cars – Triumph 2000 and Rover 2000 – were similarly priced, and were clearly aiming at the same market sector, they were remarkably different in detail. Right is a comparison of the two cars launched in 1963.

Two other statistics also bear comparison, to show how closely matched the two cars were throughout their lives :

Years on sale
    1963–77   1963–77
Total production
    248,959 **   316,962
Average annual
production (approx)
    17,800   22,600

** In addition, 80,017 V8-engined Rovers were also produced.

Even after the formation of British Leyland, there was no attempt to rationalize these two designs, or to bring them closer together. Until the day that Rover-Triumph was formed, the two designs competed just as strongly with each other as they did with other concerns. Their combined skills, however, were applied to the development of the Rover SD1 – the 3500 – which was launched in 1976.

	Rover 2000	Triumph 2000
Construction	Base unit monocoque, with bolt-on skin panels.	Conventional monocoque.
Overall length (in/mm)	178.5/4,534	173.75/4,413
Overall width (in/mm)	66/1,676	65/1,651
Overall height (in/mm)	54.75/1,391	56.0/1,422
Wheelbase (in/mm)	103.4/2,626	106.0/2,692
Unladen weight (lb/kg)	2,767/1,255	2,576/1,168
Engine	4-cylinder, overhead cam. 90bhp	6-cylinder, overhead valve. 90bhp
Transmission	4-speed, no overdrive. Optional automatic from 1966.	4-speed, optional overdrive. Optional automatic from 1964.
Suspension	IFS, De Dion rear.	IFS, IRS
Steering	Worm and roller.	Rack and pinion.
Wheels/tyres	6.50-14in radial-ply.	6.50-13in cross-ply.
Brakes	Discs at front and rear.	Discs at front, drums at rear.
UK Retail Price (January 1964)	£1,265	£1,094

This is the simple but effective MacPherson strut front suspension of the 2000, all mounted on a pressed steel cross-member. By this time Standard-Triumph were committed to rack and pinion steering for all their new models.

I think the rear suspension of the Triumph wasn't all that good. It was faulty, for two reasons. The basic idea was not bad at all, but the drive shaft splines could lock up and never let go. There was also too much swing axle effect. The angle of the semi-

trailing arms was such as to give too much effect. To improve it you would have had to alter everything. On later vehicles which were never put on sale – Lynxes for instance – we did it differently. On the other hand, I think the front suspension of the Triumph,

Technical director Harry Webster was proud of designer Mick Bunker's work on the independent rear suspension: 'That semi-trailing arm did so many jobs, yet it was a simple aluminium casting.'

For comparison, here is the independent rear suspension used on the TR4A, TR5 and TR6 models – similar in concept, but very different in component detail.

the MacPherson strut, was actually better. I think the Rover front suspension had a fault, that as you got on to full bump it went on to positive camber, so the thing suddenly understeered badly.

The road noise of the Triumph was better than the Rover – the Rover noise was a complete disaster at one stage, there's no other way of describing it.

What always impressed me about both 2000s was how good the interiors looked – particularly the Triumph, which was a

credit to the body people. Triumph had a very good bunch of people doing interiors.

Customers who tried both cars before signing orders obviously had their own dislikes, because over the years the two rival machines sold in similar, large, numbers.

Incidentally, although Rover's P6 assembly line had been moving – slowly, admittedly – since the spring of 1963, it took ages to speed up. Triumph had already built

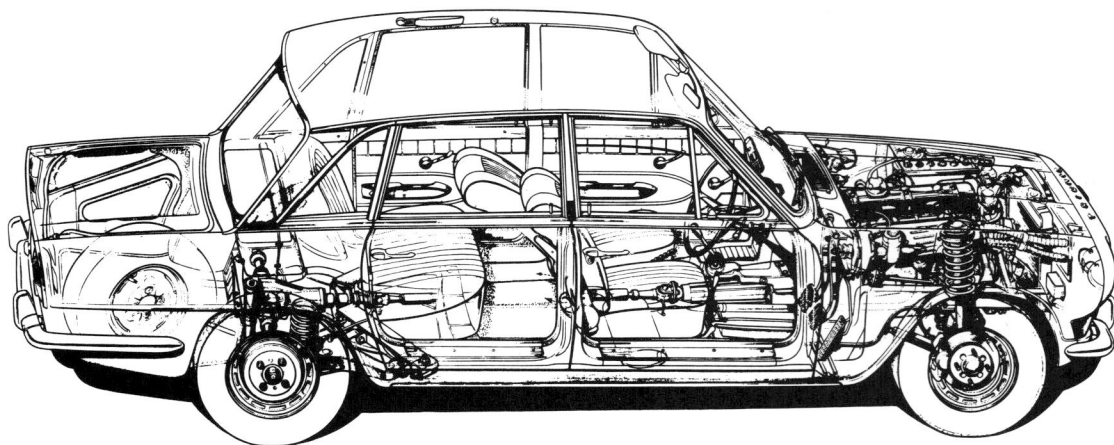

This ghosted view of the original 2000 production car shows that the tail was quite short by contemporary standards (it would be lengthened in 1969 for the Mk II model). This was a 'basic' example, without the overdrive fitted to so many of the cars.

The definitive 1,998cc engine as used in the 2000. It is fascinating to compare this with the Solex-carbed engine depicted in Chapter 1. More, much more, than the carburation had been modernized.

more 2000s by the beginning of 1964, when Triumph 2000 deliveries began, so there was no leeway to make up. In recent years I unearthed an old schedule that shows that four 2000s were built in August 1963, six more in September, and forty-five in October. Then the floodgates opened – 558 were produced in November, and no fewer than 980 followed before the end of 1963. The first 1,000-plus month was January 1964, when 1,254 examples were completed.

MEDIA REACTION

If the media ever discovered that there might also have been a Triumph 1600 as well as a Triumph 2000, nothing was ever said, and the 2000 was greeted as a great leap forward. *Autocar*, still a very discreet and supportive publication in those days, commented that:

> For some time it has been all too obvious that the faithful Standard Vanguard and its four-cylinder-engined stablemate, the Ensign, had been overtaken technically and in body styling by almost all their rivals, and that Standard-Triumph must introduce a replacement for them very soon. With the new car, to be called the Triumph 2000, the very old-established name of Standard disappears from the motoring scene partly, no doubt, because of the inevitable confusion it has caused in such contexts as standard and de luxe bodies.
>
> Replacement is a somewhat loose term, however, since this new Triumph advances a full rung up the social ladder . . . The parent Leyland Motor Corporation's philosophy behind the new car is that the smaller Standard-Triumph organization cannot easily compete with the giants – Ford, the British Motor Corporation and Vauxhall – in the lower price range, and that there is an obvious gap in the market between these and the products of, for instance, Rover and Jaguar-Daimler.

At the end of its technical analysis, *Autocar* summarized:

> All in all, the Triumph 2000 seems to have golden prospects; its modern technical specification, full equipment, roominess and pleasing proportions show no vestige of insularity, so that it should penetrate deep into export markets with the extensive backing of Standard-Triumph International and the Leyland parent.

All this was heady stuff, and it was backed up by Britain's other 'establishment' motoring weekly, *Motor*:

> With an attractive and luxuriously equipped unitary-construction body offering comfortable occasional five-seater accommodation within the easily parkable overall dimensions, the '2000' shows promise of giving just the sort of motoring many people want for the conditions of today.
>
> In short, the Triumph 2000 is a car which promises the type of motoring many people want with all the amenities they could reasonably desire.

Triumph's advertising agency, for its part, had gone right over the top, for the new model was described like this:

> The masterly new six-cylinder TRIUMPH 2000 introduces *grande luxe* motoring at a medium price.
>
> It is fast (nudging 100mph). It is very beautiful (the long low look interpreted by Michelotti). It is very quiet (a six-cylinder engine). It is a delight to drive (all-round independent suspension). It is eminently luxurious.
>
> The Triumph 2000 has been built with one thought in mind. To make motoring the civilized pleasure it should be for the driver and for his passengers. This simple aim has influenced every detail of the design of the car.

By the end of 1963 assembly of 2000s was well under way at Canley, though sales would not begin until January 1964. The commission number of the car closest to the camera – MB17193 – dates this shot in late 1964.

The advert then went on to list no fewer than twenty-seven design features.

SALES STRATEGY

All this, however, was slightly premature. When the 2000 made its debut, there were firm prices but no cars in the showrooms. Not only that, for although automatic transmission was listed as an option, no such cars would be delivered until the summer of 1964. These were the original prices quoted for the 2000. Today's 2000 enthusiasts will be intrigued :

(In those days, there were twelve pence (d) in one shilling (s), and twenty shillings in a pound. . . .)

Sales, in fact, were not scheduled to begin until 8 January 1964, so anyone out there who has a 1963-registered car probably owns an ex-factory machine. When the 2000 was first shown to the press, the company announced:

A batch of the early cars is to be sent out to representative operators throughout the country, who will report back on their experiences and troubles, if any. Where desirable and possible, retrospective

	Basic	Total (including Purchase Tax)		
2000 saloon	£905.00	£1,094	2	1
Overdrive		£ 54	7	6
Automatic transmission		£ 94	5	0
Dunlop SP (radial-ply) tyres		£ 8	9	2

modifications will be made, and on 8 January next, sales will start to the public.

Cynics later said that this was a much better strategy than that being employed by the British Motor Corporation at the time. Triumph, at least, were offering free cars on loan for customers to do its final development – as far as BMC was concerned, customers had to buy the cars first!

Although this was a thoughtful move, which brought great publicity to the company, only forty cars were actually loaned out, some of these going to senior Leyland and Standard-Triumph managers, suppliers, and some of the dealers. In some ways, though, this was an empty gesture, for little could be achieved in such a short period.

Because of the limited time scale involved – the first cars were loaned out in November 1963, opinions had to be collated by Christmas (less than eight weeks later), and deliveries to private customers began in January 1964 – Standard-Triumph was hoping only for good news. If anything awful had been uncovered, then deliveries would have had to be delayed while changes were made. The following, however, is a summary of what the company *said* had been learned in its 'consumer survey':

Using forty different cars, persons selected to take part in the test included builders, company directors, engineers, insurance agents, commercial travellers, salesmen, a company registrar, an accountant, a housewife, doctors, a brewer and representatives of the police.

Each participant was asked to award marks ranging from one to ten, according to their assessment of twelve different aspects of the car, and the averages of marks awarded are as follows: appearance, 8.97; driver comfort, 9.2; visibility, 9.2; controls, 8.7; instruments and switches, 8.24; power, 8.94; silence, 8.48; roadholding, 9.29; manoeuvrability, 9.32; fuel economy, 8.65;

passenger comfort, 9.15; luggage accommodation, 8.78.

They were also asked to comment on individual features and, naturally, there was considerable variation in opinion. The diaphragm clutch, for example, was 'too light' for some, but others found it excellent. Fuel consumption varied from 22 to 37mpg (12.8–7.6l/100km).

The main object of the survey was to discover faults which had not occurred in prototype testing, but which showed up on the first examples of production line manufacture. As a result, the throttle operating mechanism has been completely redesigned; the radiator grille has been modified to overcome an annoying wind whistle, and disc brake pads have been reshaped to avoid squeal.

Standard-Triumph's new chairman, Donald Stokes, referred obliquely to BMC's growing reputation in this summary: 'We are delighted that we decided to pioneer this form of testing. We have done our production testing on the *potential* customer and not, as so often happens, on the customer who has already paid his money.'

January 1964: On Sale at Last

By the time deliveries began in January 1964, production at Canley was already up to 350 cars a week (with 400 a week planned for later in the year), most British dealers already had cars in stock, and 2000s were also on their way to several overseas territories. The first CKD (completely knocked down) packs had already gone to Belgium, where Triumph 2000 assembly was also planned from the Malines plant.

Most people had come to the same conclusion – that the 2000 was a refined and comfortable car, if not exciting to drive. Compared with any previous Standard or Triumph, it was quieter, smoother, and

Lord Stokes

From 1961, when Leyland took control of Standard-Triumph, to 1975, when British Leyland was nationalized, Donald Gresham Stokes was the most important personality dominating the Triumph marque. After Leyland's long-time chief, Sir Henry Spurrier, had died in 1963, Stokes was the driving force behind a series of mergers that led to the formation of British Leyland.

Donald Stokes joined Leyland as an apprentice, rose to become a young and vigorous lieutenant-colonel in the Army during the Second World War, returned to Leyland to concentrate on export sales, and by 1953 was on the Leyland board of directors.

When Leyland took over Standard-Triumph in 1961 Stokes joined the board of directors, and after the mass dismissal of old-guard Standard-Triumph bosses in September 1961 he became Standard-Triumph's sales director under Stanley Markland.

In those early months it was Markland and Stokes who gave the go-ahead for the Triumph 2000 project. After Stanley Markland resigned at the end of 1963, Stokes became chairman of Standard-Triumph, presiding over the successful introduction and career of the original 2000.

Donald Stokes was Standard-Triumph's sales director when the 2000 was conceived, and chairman when it went on sale. He later went on to become the boss of the entire British Leyland organization.

Knighted in 1965, by which time he was deputy chairman and managing director of the Leyland Group, he masterminded the tie-up with Rover (1966–7), and the formation of British Leyland (January 1968), as well as being the hands-on boss of the Standard-Triumph operation. Even after he became British Leyland's chairman (from the end of 1968) and was ennobled (in 1969), he still found time to involve himself in the evolution of Triumph, which was very much the favoured marque in the British Leyland era before nationalization.

Lord Stokes always took pride in the way Leyland had speedily returned Standard-Triumph to profit in the early 1960s, and in the way that his very compact team (led by George Turnbull and Harry Webster) managed to develop so many exciting new cars in the 1960s.

Because he had always treated Triumph as a favourite son in the Leyland combine, there was no surprise that Turnbull and Webster were both asked to join him at the Longbridge headquarters of British Leyland to sort out Austin-Morris in 1968. It was not surprising, either, that Triumph always took precedence over MG in this period. If Lord Stokes had not been British Leyland's chairman, I doubt if the Triumph 2000 would have been allowed to carry on for so long, in direct competition with the Rover 2000.

After British Leyland was nationalized in 1975, Lord Stokes stayed on as honorary president for two years, but then left the company he had controlled for so long and took on other interests. By the 1990s he was living in Bournemouth, and showed no sign of retiring completely from business.

Performance	Triumph 2000 (*Autocar*)	Triumph 2000 (*Motor*)	Standard Vanguard Six	Rover 2000
Top speed (mph/km/h)	92.8/149.3	97.6/157.0	83.5/134.4	102.5/164.9
Acceleration :				
0-30mph/50km/h (sec)	4.0	4.1	5.0	4.2
0-60mph/100km/h (sec)	14.1	13.6	19.7	15.1
0-80mph/125km/h (sec)	29.9	26.8	–	29.4
Standing start quarter-mile (sec)	19.4	19.4	21.5	19.9
Direct top gear acceleration:				
30-50mph/50–80km/h (sec)	9.2	7.7	10.7	10.9
50-70mph/80–110km/h (sec)	10.8	9.8	16.2	13.6
70-90mph/110–145km/h (sec)	23.4	20.4	–	23.0
Overall fuel consumption (mpg/l/100km)	24.5/11.5	23.5/12.0	20.0/14.1	24.0/11.8

better handling with a good ride. It was, in fact, well on the way to challenging Rover at their own game, and those of us who were around in Coventry at the time also heard that Rootes were thoroughly worried about its impact on their Humber models.

Early in 1964 the specialist press gave the new car a good send-off. *Autocar* noted of its overdrive-equipped 2000 (2762 KV) that:

It is more sophisticated, more refined and more expensive than its predecessor . . . The handling and ride comfort attain a very high standard . . . most occupants commented on the very low level of noise . . . the body is obviously rigid and the steering responsive, so that the car should find favour in the less-developed overseas territories . . . there is quite a degree of understeer. [It summarized] The Triumph 2000 will not cause a flutter of excitement if one is looking for scintillating performance, but the more it is driven the more one likes it, particularly the good suspension and freedom from noise which contribute to an overall high degree of comfort . . .

Motor used a sister car (2761 KV, also with overdrive) and drew similar conclusions:

In smoothness, quietness and flexibility it sets a pattern typical of the car, with one reservation. To get the best out of it, especially if you drive fast, you must spend £54 more for the Laycock de Normanville overdrive . . . The ride and roadholding are all that one might expect from a good modern design with independent suspension all round, and the road noise level is very low: with a particularly solid rattle-free body this contributes to an unusual feeling of insulation from shock and strain, even on very rough surfaces . . . general standard is high, and the driving position makes the car feel very safe. It is.

Motor's car was significantly faster than the *Autocar* test car, but both were a lot quicker than the Standard Vanguard Six had ever been. It is also fascinating to compare Triumph 2000 performance with that of the Rover 2000 (*see* table above).

Motor's 2000 must have been slightly more powerful than the *Autocar* car, probably because it was a higher-mileage car (the test appeared eight weeks later), but there was really little in it.

Except in its higher top speed, the Rover 2000 was slower in all respects, and even though it had higher overall gearing, it was a lot heavier and was therefore no more economical. The Rover was also considerably more expensive at launch – £1,264 compared with £1,094 – and had a significantly smaller passenger cabin.

As to the Standard Vanguard Six, one only had to look at its performance, and the (veiled) criticism in contemporary magazine road tests to realize how much of an improvement the Triumph actually was. Looking back, it is now easy to see why Standard Vanguard Six sales had almost dried up by 1963, for less than 1,000 cars were built in that final year.

FILLING OUT THE RANGE

The new Michelotti-styled Triumph got off to a good start, and never flagged. The initial 350 cars being produced every week was never enough, and would have to be doubled in future years; there was usually a waiting list for deliveries. At no time in the next few years did Triumph ever have fields full of cars, rusting away and awaiting orders.

In the first year there was a lot of healthy rivalry between Triumph and Rover, for their cars were similar in so many ways. It was generally agreed that the Triumph was more refined, smoother and more spacious, whereas the Rover was technically more advanced and had better roadholding. The Rover was 10 per cent more expensive than the overdrive-equipped Triumph – but was that extra value? This lounge-bar discussion went on for years – and while no one suffered,

there was no emphatic winner either. Both cars were built for fourteen years, and although many more six-cylinder Triumphs were sold than four-cylinder Rovers – 316,962 vs 248,959 – Rover also built more than 80,000 V8-engined derivatives. Total annual production rates (of all types) were always remarkably similar.

Right from the start, a high proportion of cars were ordered with the optional Laycock overdrive transmission, and there was certainly no complaint about performance or operating economy. Complaints, if any, centred on the lack of automatic transmission on the earliest batch of production cars (deliveries did not begin until the summer of 1964), and there was lively disagreement about the instrument panel styling; the white faced instruments were not liked by everyone.

Cars for North America

Even at this stage Triumph thought there was a good potential market for the 2000 in North America, though the first deliveries were not made until late 1965. Triumph, at this stage, was on the crest of a wave in North America, where its sports cars – Spitfire and TR4 (soon to be TR4A) – were great successes. The company should have noted, however, that the Vitesse ('Sports Six', as it was known in the USA) was only selling slowly . . .

It might have been relatively exclusive, and it was certainly well made, but in the United States the 2000's problem was that it was neither fast enough, large enough, nor cheap enough to corner a worthwhile market. It was not as if Triumph had the reputation of a Jaguar. In 1965, for instance, a middle-price American family saloon such as a Pontiac Tempest sold for around $2,400 and had a choice of 140bhp six-cylinder or even 285bhp V8 engines: such a car had a 115in (2,921mm) wheelbase, a range of optional equipment, along with coupé, estate and convertible options. The 2000, on the other hand, was even smaller than the latest American 'compacts', with a mere 90bhp it was less powerful than any of them, and prices started at $2,995, with options and delivery charges to be added.

When *Road & Track*, one of the USA's most respected motoring magazines, tested a 2000 in California in October 1965, the new model was described as: '. . . a full-size imported sedan with an excellent 2-litre six-cyl engine, independent suspension all round and a pleasing appearance. A nice car.' This, by the way, was even though the importers had provided a car without overdrive: testers commented that '. . . though we had no complaint to make, we believe the overdrive would make the car more pleasant for normal use.' It was summarized as:

> All in all, it's an interesting, well-engineered car and the samples we have seen seem well put together. It is a bit pricey in the American market and the near $3,500 (delivered) price tag puts it into a tough marketing bracket. It is a distinctive car, however, and has features to which you can become attached. In our opinion it excellently fills the niche it was aimed at – 'for the man who wants more than bread alone, but can't afford caviare.

By the late 1960s Triumph-USA had virtually abandoned the 2000 to its fate. Immediately after the formation of British Leyland Bruce McWilliams came over from New York to urge them to make the 2000 look 'more American', which in his terms meant putting more glitz on to the body. One styling studio proposal, showing mock wood inserts along the flanks – 'Woodie' style – is best left unseen.

The 2000 Estate

Although the 2000 Estate did not appear until October 1965, styling work on the car had started in 1962, even as the first proto-type saloons took to the road. As ever, it would take time to convert a good idea into an acceptable shape, and in all cases it would have to be done without spending too much capital on tooling costs. For all the usual reasons, this meant that the same floor pan, front end, screen and doors would have to be retained.

With sketches supplied by Michelotti, but with much input from Standard-Triumph's Arthur Ballard, the first full-sized mock-up was completed in October 1962. However, as the pictures show, this was not yet a very graceful car, for the tail was too sharply cut off, there was a pronounced lip over the tail-gate, and there was too much angularity around the rear quarters.

Although the original style progressed only to the full-size mock-up stage, the intention was clearly to have a two-piece tailgate, with the lower half hinged at its base (the Vanguard III Estate had used this arrangement), or even to arrange for a one-piece tailgate to have glass that could be retracted downwards into the pressings – the latter arrangement was very popular in the USA at the time.

After a complete rethink, the new rear end was reshaped *and* re-engineered, so that there would be a large, one-piece, lift-up tail-gate with hinges in the roof. The rear of the cabin was made more rounded and somehow more graceful than before. Not only that, but the saloon car's tail-lamps were retained – which was no small consideration when tooling costs and lead times were considered.

Six months later, in May 1963, the shape we now know so well had evolved, though some months would elapse before a prototype could be completed. At the time, indeed, its problem was that it began to look like a fine car looking for a market that was not really there. Even after the 2000 had been launched to great acclaim, Donald Stokes's sales specialists thought that only about 2,000 Estates could be sold in every year.

The first attempt to produce a 2000 estate car resulted in this mock-up, which was not at all as graceful as intended. There was a great deal of 'Detroit-influence' around the tail, for the rear glass was intended to be wound down into the lower tailgate, which would let down from hinges at the base. It was not liked, and was then reworked into . . .

When approached to produce Estate shells at Swindon, Pressed Steel asked for £400,000 to produce new pressings, jigs and fixtures, a figure the company considered to be too high. After that setback, all work stopped for almost a year. Then, legend has it, one day a Triumph manager drove past the Carbodies factory in Holyhead Road, Coventry, to see a transporter-load of large Humber Estate car shells leaving the front gates.

Humber Estates started life as partially completed saloon shells, built by Rootes' own subsidiary, BLSP, in Acton, west London, after which they were then sent to Coventry for conversion and completion by Carbodies. If Carbodies, which was only two miles away from Canley, could do this for Rootes, would they do it for Standard-Triumph?

Negotiations, once started, progressed quickly. Not only was Carbodies happy to be involved in this work, but Pressed Steel (which already supplied a number of other part-complete shells to this company for completion and onward transmission to clients) was happy to do business with them. Because Standard-Triumph also knew that the big Humber Estate did not have long to live (it would only last until March 1967, and Carbodies knew they knew!), Donald Stokes realized that Carbodies must be looking for new business to keep its factory active . . .

The deal was done this way. For this contract, Pressed Steel would partly build batches of 2000 saloon car shells at Swindon, then transport them up to Coventry where Carbodies, experienced in low-volume batch production, would complete them as estate car shells. Then, still unpainted, they would be trucked four miles across the western outskirts of Coventry to Triumph's Tile Hill

. . . the definitive shape, which not only featured a one-piece, lift-up, conventional tailgate, but smoother details around the corners. Second thoughts, for once, were correct thoughts.

plant, where they would join saloons in the long march through the body preparation and paint process.

Final approval for 2000 Estate production came in September 1964 and the very first production cars were produced a year later. Public launch came on the eve of the 1965 Earls Court Motor Show, and it is a measure of the way that embargoes were not broken in those days that it came as a complete surprise to the public.

Structurally, the Estate's shell was closely based on the saloon, for except for the placing of damper pillars the layout of the semi-trailing arm independent rear suspension meant that this did not intrude on the estate car's floor. The saloon's front *and* rear passenger doors were retained, as were the rear wing pressings and the tail lamps. The roof pressing was new, as were the rear quarters, and the counter-balanced tailgate lifted up so far that it needed tall people to reach it to pull it down again!

The petrol tank (normally behind the rear seats on saloons) had to be redesigned, and a new 11.5 gallon (52 litre) container lived under the floor, alongside the spare wheel. Apart from re-aligning the exhaust tail pipe, the only other technical change was that the rear springs were stiffened up, and 175-13in Dunlop SP radial-ply tyres were standardized: they remained optional, for the time being, on 2000 saloons. As usual, overdrive and automatic transmission were both optional.

As in other estate car layouts, the rear seat cushion could be folded forward and the rear seat squab folded down to maximize loading space. The loading surface in this configuration was more than 5ft (1.5m) long, was fully carpeted, and had loading/rubbing strips to protect the carpets.

Although the 2000 Estate was a full generation ahead of the last of the Vanguard Estates, a space comparison is instructive. Although the roof of the 2000 was 5.5in (140mm) lower, there was the same amount of headroom above the loading floor and the

Carbodies Ltd

What had originally been an independent builder of motor car bodies in the 1920s eventually became a part of the BSA Group (which also owned Daimler-Lanchester) in 1954. By that time the Holyhead Road factory, which was cheek by jowl with the Coventry–Nuneaton railway, and which faced the modern Alvis plant, was not only noted for building bodies of all types for Daimler, but for producing convertible and estate car coachwork for such firms as Ford and the Rootes Group, and for completely assembling taxicabs for Austin.

Carbodies produced Triumph 2000/2500/2.5PI estate car monocoques for Standard-Triumph, by accepting partly built shells from Pressed Steel, completing them, and shipping them across town to the Canley assembly hall. This was the only major contract Carbodies ever carried out for Standard-Triumph.

After the 2000 contract had been completed, and after other private car work also dried up, Carbodies concentrated on producing taxicabs. For some years now the business has been owned by London Taxis International (a subsidiary of the Manganese Bronze group), which makes Nissan-engined taxicabs, and all traces of the old bodybuilding activities have been erased. The main building facing Coventry's Holyhead Road has now been sold, for Carbodies is a much smaller business in the 1990s.

size of the rear door was considerably larger. The length of the loading floors, with rear seats folded down, was nearly identical.

Those were the days when I was much involved in rally coverage for *Autocar*. For those occasions, along with a photographer, I always tried to borrow suitable machines so that we could grab sleep at the side of the road if necessary. More than once I was lucky enough to use a 2000 Estate – and that flat loading floor, well padded with a sleeping bag, was extremely comfortable.

By this time all 2000s – saloons and estates – were being built with larger-capacity Lockheed vacuum servos, and the range continued into 1966, more competitive than before. British retail prices for 1966 were:

Saloon	£1,119
Estate	£1,373
Automatic transmission (extra)	£ 95

Once again, Triumph was ready to start delivering cars as soon as the new model had been launched. The first three estate cars rolled down the line at Canley in September 1965, twenty-seven followed in October, ninety-three in November, and by the spring of 1966 about fifty Estates were being produced every week.

Once the new Estate was in full production and sales began, Triumph's sales department's earlier forecasts proved to be extremely accurate. In 1966 – the peak year for 2000 Mk I production – a total of 2,194 estates were built (compared with 21,237 saloons), and 1,672 were built in the following year.

The 2000's market was now well and truly established. Although it was by no means the most numerous of Standard-Triumph models – the Herald easily held that title – it had rapidly built up a prestige reputation, and Triumph surveys showed that it was reaching exactly the clientele the Vanguard Six had always missed – doctors, architects, solicitors, young businessmen and accountants among them.

MID-LIFE IMPROVEMENTS

Although Triumph was happy to leave the 2000 range's mechanical specification undisturbed for some time, there was always the

As loved by a private owner in the 1990s, this 1967–8 2000 has had various extras added, including wing mirrors and an extra driving light.

Nowadays, perhaps, we might call the original 2000 a little over-decorated, with that large badge and the overriders, but this was a very smooth nose for its period.

Consider the detail study that Michelotti and Triumph's Leslie Moore put into the front corner of a 2000. This, by the way, is really the 'Mark 1½' model produced towards the end of the model's life, complete with rubber face overriders and a more complex facia and interior.

incentive to make a good car look and feel even better. From October 1966, therefore, the cars were treated to restyled seating, new instruments and ventilation, and changes to the automatic transmission.

The original black-on-white instrument style, always controversial, was dropped in favour of a more classic white-on-black installation, and at the same time swivelling eyeball air vents and an electric clock were added to the facia. Through-flow ventilation was ensured by stale-air outlets at the rear of the cabin, while the new seats were faced with perforated leather.

Mechanically the only changes were to the Borg Warner automatic transmission arrangement, now given a D-2-1 selector

For the original 2000, Triumph chose to use a developed version of the shield that was currently used on TR models, though larger and with a more ornate surround. Each of the T R I U M P H letters was individually fixed to the nose.

The '2000' badge was fixed to the centre of the rear bumper from late 1966, the only indication of the engine size on this particular version of the car.

Triumph's 'world' badge had become familiar on TR sports cars, and was used in the middle of the 2000's wheel covers.

Here is an oddity, which proves that later owners sometimes go their own way, ignoring originality. The 'Mk 1½' 2000, introduced in October 1966. This type of facia, complete with face-level ventilation, should really have black-faced instruments with white lettering, but owner Ron Field clearly prefers the original white face/black lettering dials instead.

81

Although 1966–9 2000s had exactly the same sheet metal style as before, their badging was altered to include a '2000' badge on the rear bumper itself, and the overriders were given rubber inserts.

quadrant, but no lock-up system, while alternative spare wheel mounting positions – upright or on the boot floor – were provided. All this was achieved with a price increase of a mere £60 (to £1,198) for the saloons.

Fastback 2000GT – A Good Idea But . . . .

In the meantime, other 2000 variants had been considered, built up as prototypes, then discarded. Well before the 2000 saloon was ready for launch, Harry Webster and his colleagues had considered the derivatives that might follow. A convertible version of the car was mentioned at a board meeting in October 1963, but as the sales department thought it could only sell 1,000 cars a year (a figure raised to 2,500 a year in the weeks which followed) that idea was speedily dropped. Years later, after a great deal of mind-changing, and in an entirely different form, this car would reappear as the Stag (*see* Chapter 11).

The idea of a 2000GT, a five-door car with fastback styling, was treated much more seriously. First considered in 1963, it pro-

gressed as far as a mock-up in the spring of 1964, and Harry Webster then briefed the board of directors in April of that year.

By this time Webster's engineers, helped along by John Lloyd's development team, were becoming truly adept at mix-and-match projects. For this application, therefore, work which had already been done on the six-cylinder engine, with a view to fitting it into the TR4/TR4A structure, was reviewed for a saloon application. An entirely new cylinder head casting was being developed – experience with highly-tuned SC engines for the Spitfire competition programme, and for the rallying 2000s, would all contribute – and instead of 90bhp, between 105 and 110bhp could easily be provided.

Although the 2000GT was to have been based on the standard 2000 saloon car monocoque, it was always meant to have a fastback tail. The pictures of a styling version, which date from March 1964, tell their own story. Close inspection shows that this mock-up had different roof/tail treatments on each side, and that it *was* a mock-up, no more and no less. I am confident that a

The fastback that never was! In 1963–4 there was a serious proposal to build a fastback 2000GT model, but it was eventually cancelled in favour of the estate car body style, for there was only enough available investment for one new body derivative. The actual car shown is a 2000 saloon to which the styling department has mocked up a fastback tail. Two different detail treatments – one at each side – were being studied at this time.

normal four-door saloon shell was still hidden under those new panels.

On either side the doors and side windows were standard, as was the profile of the rear wings, and across the tail. New rear quarter windows would have been specified, the roof panel would have had to be altered behind the line of the rear doors, and there was a completely fresh look to the rear of the cabin.

Notice, too, that according to the details of this mock-up there was to have been a full size hatchback in this layout, and this was well before Renault showed that it had 're-invented' the hatchback theme with its 16. However, I cannot believe that at that stage any serious work had been done on

engineering what would have been a very complex hinging panel.

Assuming that this hatchback had to be hinged at the roof, behind the seats, it would have been very large and heavy (area for area, car glass is heavier than sheet steel), and there would have been a complex pressings profile that would have been a nightmare for production engineers to work out. All in all, I can see that the difficulties would have argued for themselves if prototypes had been built.

In the end, in fact, the project got very little further, and little running of a 'GT' engine was ever carried out in test cars. With the 2000 Estate and the new front-drive 1300 coming along, with the new TR4A nearing

This 1967–8 2000 looks immaculate, even though the pictures were taken when it was more than twenty-six years old. The body shells last well, and still look good from any angle.

production, and with the GT6 reaching the tooling-up stage, the 2000GT would have had to fight hard to gain priority. Harry Webster reported all this to his board in September 1964. At the same time that he asked for final approval for the estate, he asked for the 2000GT to be deferred. This was a polite way of suggesting that it should be cancelled, and he got his way.

There would be no more high-performance 2000s until 1967, when the first 2.5PI prototypes were built.

MATURITY

By 1967 Triumph was already planning major changes to the 2000 family, so very few late-life improvements were ever made to the original Mk I-shape cars. Engineering, in any case, was becoming progressively more and more absorbed by the Stag programme, which leaned heavily on the 2000's team of engineers for its expertise.

Between 1967 and 1969, though, two major engine changes were considered. One was a fuel-injected six-cylinder unit, which

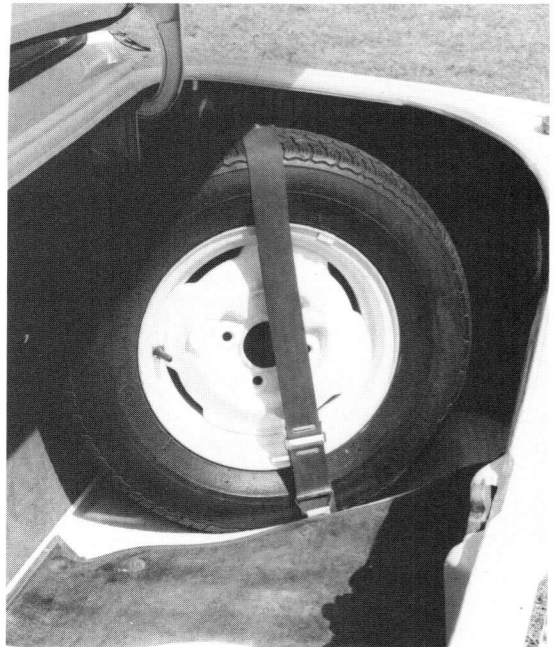

In the original 2000, the spare wheel was always mounted in the side of the boot compartment like this, but from the autumn of 1966 it was also possible to mount it flat on the floor instead.

The six-cylinder engine of the 2000 was canted over by 10 degrees towards the right side (offside in the UK) of the engine bay, to allow clearance between engine accessories and the battery tray on the other side of the compartment. Note the brake servo on the bulkhead, and the carburettor air intake pipes leading in from the nose.

eventually went ahead as the first of the 2.5PIs, but another – the use of the Stag's new V8, did not get very far.

Although confidential product plans, dated 1966, optimistically forecast that the V8 engine would become available by the end of 1968, for use in the saloon *and* the Stag shells, this did not progress much further.

Triumph's newly appointed sales director, Lyndon Mills, ran a PE158 2.5-litre V8-engined 2000 for some time, and apparently loved every minute of this, but no amount of his lobbying could convert this from 'good idea' to 'production car'.

Spen King, who took over from Harry Webster as Triumph's engineering director in the spring of 1968, confirms that the absorption of Rover into Leyland in 1966–7, effectively killed off any idea of producing a V8-engined 2000. Now that Rover and Triumph had been gathered into the same conglomerate, and some rudimentary form of product planning was already taking place, it would have been too obvious a competitor for the Rover 3500 of the period.

Even so, the Triumph 2000 continued to sell astonishing well, and usually matched the big rival – the Rover 2000 – head for head. The professional class that had been targeted was happy to buy, and buy again, a

By the late 1960s the first compulsory anti-emission devices were being fitted to the 2000. The Zenith-Stromberg carburettors, originally designed by Standard-Triumph for their own use, had been taken up by many other manufacturers in the mid-1960s.

car that looked good, behaved impeccably, and was more capable than any previous large Standard-Triumph model.

In 1963 Donald Stokes and Stanley Markland had forecast that up to 20,000 cars would be sold every year. As these figures (for the complete 2000 range) show (*right*), that was uncannily accurate.

By then, however, the range had been expanded, for great things had been achieved with the six-cylinder engine. When the new TR5 sports car appeared in October 1967, not only did it use an enlarged version of the 'six', but fuel injection had been fitted.

Year	Production
1963	1,685
1964	18,490
1965	19,087
1966	23,431
1967	19,820
1968	22,242
1969 (To September, when Mk II production began)	17,229

Only a year later Triumph did the obvious, mating the new engine with the saloon car body – and the result was the 2.5PI.

5   2.5PI – the Fuel-Injected Pioneer

I have already touched on Standard-Triumph's desire to develop a more powerful version of the 2000 model. In standard condition, and with only 90bhp, the 2-litre car's top speed was well under 100mph (160km/h), much slower than the Rover 2000. Worse, cars like Ford's latest Zodiacs and Vauxhall's Crestas were not only faster, but cheaper too.

As related, as early as 1963/4 much thought had been given to building a 2000GT hatchback, which would have had a tuned-up engine with between 105–110bhp, while the proposed 2000TS saloon would have shared the same unit. Nothing came of these proposals, principally because the higher-power engine was not nearly as flexible as John Lloyd's development engineers wanted – their masters from Leyland had insisted on the need for low-speed torque and pulling power in all new Triumph family car models.

When the decision was made to fit the new generation of TR sports cars (TR4B, or Wasp, as it was originally known, TR5 or TR250 as they went public) with the six-cylinder engine, the need for more power *and* torque became urgent. There was no question of a brand-new engine being designed, and in any case the transfer line tooling was still to be shared with that of all the four-cylinder SC engines, so all improvements to the six-cylinder engine would have to be winkled out of the same basic package. The breakthrough came in 1965 with the development of two interrelated features – a new type of cylinder head, and a long-throw crankshaft/cylinder block combination.

Because there was no scope for the cylinder bore to be increased, the existing 'six' was given a new and longer stroke dimension of 95mm instead of 76mm, which increased the capacity from 1,998cc to 2,498cc. To accommodate the extra 9.5mm 'throw' of the crank shaft, and the extra angularity developed by connecting rods in their rotation, the walls of the crank-case had to be spread slightly, so a new cylinder block casting was required. At the same time crankshaft bearing sizes were also increased. Fortunately, investment in modified casting moulds was relatively cheap, and the all-important cylinder centre machine tools did not have to be disturbed.

The new cylinder head, known as the full-width type, not only restored a more sensible way of attaching the manifolds, eliminating the possibility of blown exhaust gaskets, but it also had better-shaped ports and combustion chambers, like those of the very latest Spitfires. First used on the 2000 race car of 1966 (but diplomatically never mentioned by the factory at the time), it was first seen in public on the TR5PI in the autumn of 1967, and was then adopted for the other six-cylinder engines a year or so later.

All this work was originally driven by the need to get the projected TR4B/Wasp ready during 1967, but the opportunities for fitting similar engines to the 2000 were also apparent. However, it was some time before the product planners homed in on the model which we now know as the 2.5PI.

As described more fully in Chapter 6, from

The fuel-injected 2,498cc engine was fitted to the Triumph TR5 sports car from the autumn of 1967, when this was the installation. In this guise it had a very sporting camshaft profile, and the engine produced no less than 150bhp.

1964 to 1966 Triumph's competitions department spent much time and effort developing highly-tuned versions of the 2-litre 2000s for use in rallying and racing. Although this produced a handful of spectacular and effective cars, which sounded wonderful but were woefully unrefined, it also convinced Harry Webster that a Barb GT needed to have a larger capacity engine, with more torque. By then the 2.5-litre engine was under development, and fuel injection had come on to the scene.

In 1967 the know-alls originally jumped to the wrong conclusions. From the moment that Triumph unveiled the 2.5-litre fuel-injected TR5PI sports car, they all lined up to say: 'Well, of course, Triumph will put that engine in a saloon shortly.' Well, yes and no – we now recognize the 2.5PI as a lasting success, but approval for a new model of this type was by no means as easily gained as you might think.

FUEL INJECTION

Triumph first dabbled with fuel injection (PI stands for petrol injection) in 1965–6, when that dreaded phrase 'exhaust emission limitation' first cropped up. For cars to be sold in North America, from 1968 new regulations demanded low limits on hydrocarbons and carbon monoxide in engine exhausts. (By comparison with the incredibly demanding 1990s, the 'low limits' of 1968 now look almost ludicrously lax, though development engineers did not think so at the time). Engineers had never previously been forced to consider such niceties, and for a time there was panic in the ranks throughout Europe.

As Harry Webster once told me:

At that time, none of us knew anything about emissions, and it all looked very serious. We wanted to carry on selling TRs, GT6s and Spitfires in the USA, and at first we thought the only way to beat the limits would be to fit expensive hang-ons like after-burners, exhaust recirculation and catalysts. Lucas could supply all those things – they had a lot of experience with aero-engines – but we worked for months without success.

Eventually both firms decided that they were really tackling the problem from the wrong end – the exhaust end – of the engine. It was more profitable, they concluded, to see that the fuel-air mixture going in to the engines was more perfectly constituted, than to try to clean up the environmentally dirty mess that came out at the other end!

The irony was that this led to the development of petrol injection, for which the necessary components had to be incredibly carefully machined, and which automatically meant that costs were high – and then finding that this hi-tech (by 1960s standards) layout could not cope with the new requirements.

Lucas fuel injection

The first application of Lucas's mechanical fuel injection was in motorsport, on the works Jaguar D-Type race cars of 1956. Tested during 1955, the first actual racing appearance was on Mike Hawthorn's car at the Sebring 12-Hour race in March 1956. This system, however, was only an ancestor, not a direct prototype, of the systems that followed.

The first production-type application was on the Maserati 3500GTi of 1961. The original 3500GT had used Webers, the Lucas injection being adopted because it gave more power and a smoother power curve. Once again, however, the Maserati installation had little, visually, in common with that later adapted by Standard-Triumph.

Work began on the six-cylinder 2.5PI/TR5 installation in 1965–6, a prototype being used on Bill Bradley's works Triumph 2000 race car in 1966. A production-standard version, as just introduced for the TR5 PI sports car, was ready for use by Denis Hulme/ Graham Robson and Roy Fidler/Alan Taylor in 2.5PI prototypes in the 1967 RAC rally, which was then cancelled.

Although Lucas made strenuous efforts to sell this generation of fuel-injection system to other manufacturers, it was never adopted on any other engine. The last Lucas-injected Triumph 'six' was built in 1975. Lucas systems fitted to later Jaguars, Rovers and Triumphs leaned heavily on licence-produced Bosch expertise.

Without new-fangled and still-unproven controls to look after several temperature and altitude variations, the injection system was still not good enough – whereupon Triumph turned back to the existing Zenith-Stromberg carburettors, finding that these could do the job after all . . .

It is now well known that the company had already used the Lucas installation on the

Bill Bradley 2000 race car (described in more detail in Chapter 6), which had a 2-litre engine, and that they then progressed to tailoring it to a high output version of the 2.5-litre road car. This decision, however, was taken at a very late stage, only a matter of weeks before prototypes were entered for the RAC rally, in which one of the cars was to have been driven by F1 World Champion Denny Hulme, and me. This saga is also detailed in Chapter 8.

In Britain Lucas were certainly the resident expert in fuel injection systems. Lucas had first become involved with injection (petrol injection) during the Second World War, when British aero-engine manufacturers like Rolls-Royce began searching for more precise ways of feeding fuel to their highly rated engines. Development then continued for racing car units, the first public installation being seen on the 'works' D-Type Jaguar Le Mans cars of 1956.

Then came the long search for a production-car contract, which took Lucas another decade to achieve. Systems fitted to race cars did not count nor, because of the limited numbers built, did the Lucas injected six-cylinder engines fitted to Maserati 3500GTi, Sebring and Mistral models. What was called the 'Mk II' road-car system did not break cover until 1966, when it was shown on the Lucas stand at the Earls Court Motor Show.

Extra Horsepower, and Torque

When I wrote the description of the new TR5 for *Autocar* in October 1967, I pointed out that the Lucas injection only accounted for the last 5–10bhp, but that: 'its greatest advantages are that mixture can be controlled to such fine limits, and in such varied conditions, that a camshaft with much more extreme timing could be proposed while retaining docility and a satisfactory idle'. Or so I was assured at the time; it did not take

Triumph long to acknowledge that some examples of the TR5PI had a very lumpy idle indeed, and a modicum of detuning was carried out in due course.

Even before the TR5PI was unveiled, work had already started on the obvious transplant – Lucas-injected engine into 2000 body shell. The first official fuel-injected prototype carried the experimental chassis number of X768, and was registered LRW449F in the autumn of 1967, this being speedily followed by seven further prototypes, X769 to X775 inclusive. Although some cars were built with fuel-injected *2-litre* engines (such cars are listed as 2000PI models in the experimental register), this version of the engine was soon abandoned.

Development of the new high-performance car was completed in double-quick time, for Triumph was not proposing to put it on sale in the USA (which eliminated all the burgeoning exhaust emission proving work at a stroke), and lengthy pavé testing of the unchanged body shell was not needed.

In any case, once the fuel-injected TR5PI had been announced, and especially after the two special cars had been prepared for the 1967 RAC rally, Triumph enthusiasts began to forecast that a production car would soon be available, so Triumph dealers were soon under pressure from customers, wanting to know when the TR5-engined 2000 would be announced. In the end, they had to wait a year – a year in which there had been monumental corporate upheavals.

THE FORMATION OF BRITISH LEYLAND

Between 1966 and 1968 the Leyland Motor Corporation expanded enormously, gathering the Rover, BMC and Jaguar Groups into its empire. This led to the formation of British Leyland, which had untold consequences on Triumph in the years that

> ## British Leyland – founding a colossus
>
> After Leyland Motors took control of Standard-Triumph in 1961, they needed time to consolidate, to sort out the financial mess they had inherited, and to begin planning strategically for the future. In 1962 Leyland also absorbed the London-based bus-making company ACV (which made AEC buses), but no more acquisitions followed for three years. The chance to take a government-backed stake in Rootes (where Chrysler was already a major shareholder) was turned down, and tentative talks with BMC got nowhere. Leyland also talked to Jaguar in 1965, but Sir William Lyons was not then ready to give up his independence.
>
> Once Standard-Triumph became profitable again (*very* profitable by the mid-1960s), Leyland looked around for its next move. It came in 1966–7 when Rover (which had recently absorbed Alvis) merged with Leyland-Triumph. The potential of merging Rover's car-making expertise with that of Standard-Triumph was obvious, but little was done about this until the early 1970s.
>
> In the meantime BMC (whose major marques were Austin, Morris, MG, Riley and Wolseley) had got together with Jaguar (who also owned Daimler, Guy and Coventry-Climax) in 1965. The new holding company, British Motoring Holdings (BMH) struggled to get to grips with its amorphous shopping basket of businesses, and by 1967 was no nearer making financial sense of it all than two years earlier.
>
> With the British government (headed by Harold Wilson, and with Tony Wedgewood Benn as minister of technology) acting as marriage brokers, Leyland and BMH spent most of 1967 trying to get together, but it was not until January 1968 that the deal was finally done. The new colossus was named British Leyland, and before long it became clear that Leyland, not BMC, were the political masters.
>
> In seven short years, therefore, Triumph had moved from being a dominant marque in Standard-Triumph, to being just one of many badges in the British Leyland line-up. Worse, the rival 2000s – Triumph and Rover – found themselves almost, if not quite, on the same side, while Triumph's sports cars were owned by the same company that owned their deadly rivals, MG.
>
> For the first few years after the foundation of British Leyland, Triumph continued as an autonomous business, there being absolutely no co-operation with Rover, even though those two companies were only about 10 miles apart.
>
> By the early 1970s British Leyland had started to merge several businesses, and in 1972 it set up Rover-Triumph, with the balance of joint executive power moving subtly, but definitely, towards Rover.
>
> Separate operating companies (including the Triumph Motor Co. Ltd) were abolished in September 1972, and in May 1973 Lord Stokes announced that 'Triumph will cease to compete with Rover'. In business terms, about time too.
>
> After British Leyland was nationalized in 1975, all the car-making operations were gathered under the umbrella of Leyland Cars, and the Triumph 2000 family ended its days under that banner. By then, events had discredited this strategy and after Sir Michael Edwardes arrived as chief executive, another upheaval followed. Triumph was submerged into Jaguar-Rover-Triumph, but the last true Triumph cars were built in 1980. The so-called Triumph Acclaim was nothing more than a rebadged Honda model.

followed. The sequence of mergers was as follows :

June 1965   Rover take over Alvis.
July 1965   The British Motor Corporation (BMC) absorbs Pressed Steel. This has major

implications on the future sourcing of Triumph body supplies.

July 1966   BMC take over Jaguar (which already owned Daimler and Coventry-Climax), forming British Motor Holdings (BMH).

December 1966 Leyland make a take-over offer for Rover. This brings Triumph, Rover and Alvis into the same net.

January 1968 Leyland merge with BMH (this was, effectively, a take-over by Leyland) to form British Leyland. This means that Triumph is now owned by the same group that also controls Alvis, Austin, Austin-Healey, Daimler, Jaguar, Land Rover, MG, Morris, Riley, Vanden Plas and Wolseley.

Almost immediately British Leyland announced a far reaching study of its future new model requirements. This sounded impressive at the time; it was only later that we realized just how reluctant the planners had been to tamper with existing product plans! Nothing that Triumph had already proposed was changed for the time being – indeed, nothing significant changed for the next three years.

Pessimists immediately wondered if Triumph and Rover would have their corporate heads knocked together, and if this would affect the pedigree of either concern: it would, but not for several years. As it transpired, the Triumph 2000 and Rover 2000 ranges continued, in competition, for the next nine years, and there would be no rationalization of any type.

One immediate result, however, which had a big effect on Triumph, was that Sir Donald Stokes asked two key figures from Triumph – George Turnbull and Harry Webster – to move across to Longbridge to get a grip on the Austin-Morris side of the new business. Once there, Webster poached George Jones, Ray Bates and Stan Holmes from his old staff, thus upsetting the balance of skills remaining. To replace Webster, Sir Donald then asked Spen King to move across from Rover to Triumph to take his place. King, who was comfortably ensconced as Chief Engineer, New Vehicles at Rover, had just got the Range Rover running when the call came:

I was really enjoying my job at Solihull. I had hardly any warning of the move. It really was incredibly short. I was asked by Stokes, would I accept the job, and I had about three minutes to answer this. Then I moved in very quickly after that.

Some people probably thought that John Lloyd, who was Harry's deputy, should have had the job, but people accepted me reasonably well. Once I had moved, I had to cut all functional links with Solihull, but since Peter Wilks was there, and was a very good friend of mine, I still knew what was going on.

At the time, though, the two companies still regarded each other as rivals . . .

INTO PRODUCTION

By the spring of 1968 the new fuel-injected car was almost ready, though a lot of work had been needed to make the engine more flexible. In February Harry Webster had told his board that the introduction of the fuel-injected car would have to be delayed, as the unit was not ready for a saloon car introduction.

Taking over from him, Spen King discovered that there was still a serious problem with high-pressure fuel pump noise: 'We had to shift the pump to the back of the vehicle, where it carried on giving trouble. It was always audible, but it had been a damned sight worse when it was up front, under the bonnet.'

Very few changes were made to the existing style – only enough to give the injected car an obvious 'identity'. Leslie Moore, who was running the styling department, could only add new wheel trims, a grille on the bonnet and black vinyl quarters 'to visually lower the car'.

Although the fuel-injected car, now titled 2.5PI, was introduced in October 1968, the very first production car – chassis no. MD1 – had been assembled in July 1968. Two pre-production machines had been built as

Spen King

Although Spen King was the technical director of Triumph for only seven years, his was the task of ushering the Mk II 2000 and 2.5PI models into production and, more important, even, seeing that the Stag Grand Tourer finally went on sale. Although King's views differed from those of Harry Webster in many ways, he was a fine engineer who was well-respected at Triumph.

Charles Spencer King's mother was Rover chairman Spencer Wilks's sister, which made those two famous Rover personalities, Spencer and Maurice Wilks, his uncles. Educated at Haileybury school, he completed a war-time apprenticeship at Rolls-Royce, began to learn all about gas turbine engines at Derby, then moved to Solihull to work on the Rover gas-turbine project in 1946. The first Rover gas-turbine car – JET1 – was unveiled in 1950, and two years later it was Spen King who drove it at 152mph (245km/h) in a Belgian speed test.

Taking over as Rover's head of gas-turbine research in 1952 (where he and Gordon Bashford later designed the rear-engined four-wheel-drive P3 prototype), Spen moved on to head the company's technical forward-planning team in 1959, installed the new V8 engine in many chassis in the mid-1960s, and designed the sensational mid-engined P6BS Coupé *and* the original Range Rover 4x4.

After the upheaval triggered off by the formation of British Leyland, King was drafted in to Triumph as technical director in 1968, and from 1972 was put in charge of the entire Rover-Triumph engineering effort. This was the period in which the TR7 and Rover SD1 projects were shaped – both of them with solid rear axles, for King did not believe in (as he put it), 'half-baked irs systems'. It was also the period in which Triumph fell out of love with the Lucas fuel injection, which was neither reliable nor cheap enough for their tastes.

Then, in 1975, he was moved back to Longbridge to direct the technical fortunes of the entire Leyland Cars operation (a job, which reputedly he did not enjoy very much as there was too much administration and not enough actual designing). Later, from 1979, he became deputy director of BL Technology Ltd at Gaydon, where he ended his career in 1986.

Retiring (though still consulting to other engineering firms) to his beautiful old house in rural Warwickshire, he finally found time to indulge himself more in boats, and to have a limited-edition Range Rover (the sporty CSK model) named after him in 1990.

early as May 1968, while regular assembly began in September and October. Once launched, deliveries began at once, and during 1968 around 400–500 2.5PIs were built every month.

For some time afterwards Triumph's advertising campaign majored on the existence of 'the power elite', claiming that annual output was restricted. Although there was an element of limited production in what Lucas could do at the time, there is little evidence of long waiting lists for what was a rather more expensive car than the 2000 itself.

In many ways this had been a straightforward development job, for the engine was no more than a detuned version of that used in the TR5 sports car, and there were no major style changes. Externally, the only way to pick a 2.5PI from its 2000 derivative was by the new 'Injection' badge on the bonnet, by the new badging on the tail, the black vinyl covering of the quarter panel, and by the use of TR5 dummy-Rostyle wheel covers.

The major advance, of course, was in adopting fuel injection for a production car, this actually being a UK first. Over in Europe there had been fuel-injected Mercedes-Benz models since 1959, and Peugeot had followed suit with the 404 Injection in 1964, but Triumph was definitely the pioneer in the UK. Lucas, as suppliers of the system, made much of this in their advertising:

For use in the projected 2.5PI saloon, the Lucas fuel-injected engine was slightly detuned: instead of 150bhp, as fitted to the TR5 and later the TR6, it was given a 'softer' camshaft profile, and produced 132bhp. This is the original 2.5PI installation, complete with large air cleaner, and a mass of piping! Note the plastic cooling fan. There is no overdrive on this particular installation, but many 2.5PIs did have overdrive fitted.

To identify the 2.5PI, Leslie Moore was only allowed to make minor changes to the 2000's style. One such detail was a new badge fitted to the front of the bonnet bulge. As this type of 2.5PI was only in production for a year, such badges are now rare.

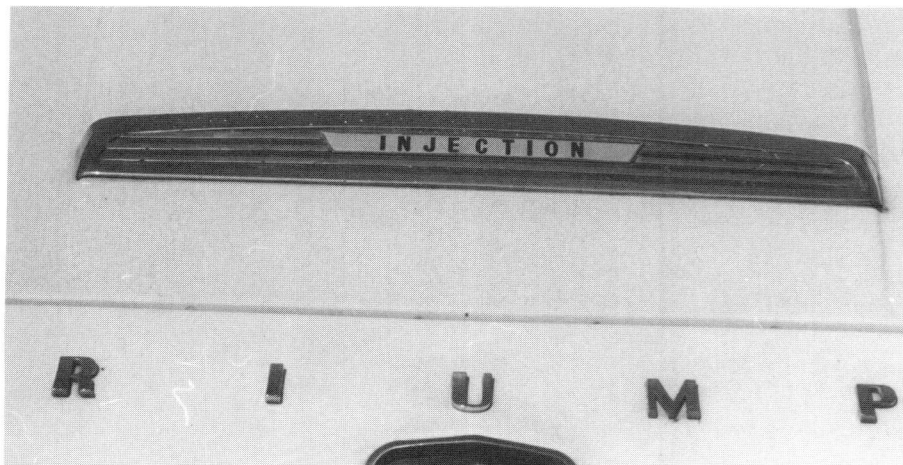

The original type of 2.5PI was available in saloon or (as here) estate car guise. This car, pictured in 1994, had been given wing mirrors and extra driving lamps. The 1969-model PI was the first big Triumph to use the dummy-Rostyle wheel trims.

From the rear, the only way to pick out the original 2.5PI was by the badging on the tail and on the rear bumper. The bumper badge, quite succinctly, proclaimed '2500' instead of '2000'. Estate car deliveries of the 2.5PI began in the spring of 1969, and were completed by the autumn of the same year, which makes this the rarest of such types.

For the first time ever, petrol injection has been fitted to a British family car. And Lucas know-how made it possible. Petrol injection is the most advanced fuel control available. And Lucas is the most advanced type of petrol injection. That's why Triumph chose Lucas for their 2.5PI. For smooth precise power and peak efficiency.

The TR5 Connection

Although the 2.5PI engine installation was visually nearly identical to that of the TR5PI sports car, it was a significantly less powerful unit. Like the TR5PI, it used the same new full-width cylinder head (which had also been adopted on Vitesse 2-litre Mk II and GT6 Mk II types), and the same inlet manifolding with its forest of injector pipes. The high-pressure vane-type fuel pump was mounted in the boot, near the fuel tank. Triumph themselves published a comparison of the engines (see table below).

Engine	Peak Power (net)	Peak Torque (net)
2.5PI	132bhp @ 5,450rpm	153lb/ft. @ 2,000rpm
TR5PI	150bhp @ 5,500rpm	164lb/ft. @ 3,500rpm
2000	90bhp @ 5,000rpm	117lb/ft. @ 2,900rpm

For the 2.5PI, Triumph's engineers wanted to provide a more flexible engine than the TR5PI, with none of the uneven idle for which the TR5PI was already becoming famous. To do this a less extreme combination of camshaft timing and lift was chosen, and there was a single outlet cast exhaust manifold.

Model	Camshaft timing (degrees)	Valve lift (inches)
2.5PI	25-65-65-25	0.336
TR5PI	35-65-65-35	0.367

A comparison of the two camshafts (see table above) shows that for the 2.5PI there was considerably less valve overlap (50 degrees around top dead centre, instead of 70 degrees for the TR5PI), and less valve lift. In fact the camshaft profile of the 2.5PI engine was exactly the same as that being used on the latest Spitfire Mk 3 four-cylinder engine – Standard-Triumph was trying very hard to rationalize on specifications and machining operations at this stage in its history.

As with the TR5PI, the compression ratio was no less than 9.5:1, which put the octane requirement of the engine perilously high (in the ecology-conscious 1990s it is not possible, for instance, to use lower-octane lead-free fuel in unmodified PI engines), and there was also a centrifugal cut-out in the Lucas distributor, set to operate at 5,800rpm. Because no rev counter was fitted to the original type, the driver could not keep visual checks, so this cut-out was fitted to keep the engines under 6,000rpm, where there was a troublesome crankshaft torsional vibration 'period'.

To match the extra power, a gearbox with unchanged ratios was fitted with high-capacity needle roller bearings, while the optional Laycock overdrive had higher oil pressure settings. The propeller shaft was a twin-tube type, with rubber inserted between the tubes, and (as with the TR5) the final drive ratio was raised to 3.45:1. This meant that with overdrive fitted, the new 2.5PI was really long-legged – 24.5mph (39.4km/h)/1,000rpm in overdrive top gear.

To match all this, the new car had ⅛in thicker brake discs, there was a different

This beautifully preserved 2000 shows off the lines conceived by Michelotti in 1961, refined at Coventry later in the same year, and first shown to the public in 1963. There is not a clashing line, feature or angle anywhere. No wonder these cars sold so well, for so long.

Michelotti's original style for Barb featured a stubby tail, but the sill for the boot was low. In many ways, the Triumph 1300 that followed two years later copied this style, while the tail lamps were always retained for use in the big estate cars.

Compare this cockpit shot with that of Mark II models, and I think most people will agree that the original 2000 facia layout was somewhat fussy. No one, however, could complain about a lack of standard equipment.

Triumph tester Gordon Birtwistle was driving this 1966 2000 for an advertising photo-shoot. This tracking sequence was meant to show the links between Triumph and Leyland. The truck cab, by the way, was also Michelotti's work.

The prototype Triumph 1600 (note the vertical-slatted grille) was given a 2000 engine for 1964, and handed over to the competitions department as a service car. Along with a trailer, and, overloaded with tyres, it is ready to follow the route of the 1964 Spa–Sofia–Liège rally.

When shaping the Innsbruck (Mk II) style, Michelotti was encouraged to make the tail longer. This not only looked good, but provided more stowage space in the boot. The back looked similar to, but was quite different from, that of the Stag.

The Mk II models had entirely different noses from the Mk Is. Original cars were rounded with a bonnet bulge, and with headlamps recessed to each side; Mk IIs like this had a full-width sweep, and the headlamps almost flush to the grille bars.

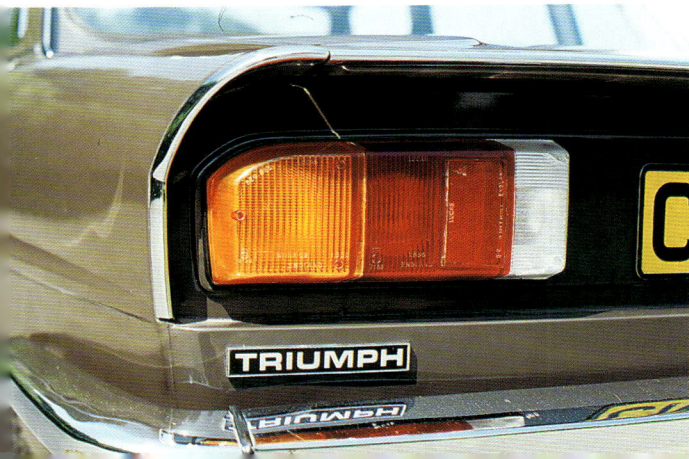

Rear tail lamp/indicator clusters on Mk II models were similar to, but not the same as, those of the Stag. Should British Leyland have insisted on more standardization? Stylists resisted, but would the public even have noticed?

From this angle only it is easy to see where the Mk II nose style was grafted on to the original car's sweep of front wings and bonnet profile. The bonnet pressing, the front wings, indeed.the whole front end ahead of the wheels, were pressed on new tooling from late 1969 onwards. (Below) For the Mk II models, Triumph's body designers produced a series of sumptuously trimmed and equipped interiors. This particular car is a 2.5PI, complete with sports steering wheel, and matched rev counter and speedometer. The wood veneer lasts very well, and is well worth restoring today.

Introduced in 1975, the 2500S was really a less powerful replacement for the troublesome 2.5PI. Complete with a 106bhp 2,498cc engine, firmed up suspension and cast alloy wheels, it was a brisk and capable executive saloon.

Not *easy to restore or prepare for concours – for cleaning up those grille bars and making good the lighting details takes a lot of time. This 2500S looks right, even twenty years on.*

For the 1975–7 cars, there was a new and more curvaceous inlet manifold, and on cars like this 2500S there were SU HS6 (1¾in throat) carburettors. This is not a concours 2500S, but a well-loved and regularly used example of the mid-1990s.

From this angle, it is difficult to spot that this 2500S is actually an estate. The wing mirrors on this car are accessory items.

On all the big Triumph estate cars, the rear loading door was a lift-up item, very difficult to press and trim (and keep watertight), but neatly integrated into the otherwise little-changed structure. The towing hitch on this car was fitted by the owner, not provided by the factory.

Although Barb and Innsbruck estate cars weighed a little more than the saloons, their performance was only slightly impaired. Overall the bulk was the same and the trim and furnishings were up to saloon standards; they were, and still are, very popular.

Six-cylinder Triumph engine bays were always full, but by tilting the engine slightly towards the right side (closest to the camera) there was space for an alternator and the battery tray. To fit the longer inlet manifolds in the mid 1970s, a very slim air box had to be packed in close to the turret pressings.

Using a short-wheelbase version of the 2000/2.5 platform, Michelotti produced an elegant 2+2-seater cabriolet style for the Stag. This is a later model, with the cast alloy wheels that were also used on the 2500S saloon. Note the family-likeness tail end.

The Stag's distinctive T-bar was necessary to eliminate the scuttle shake that afflicted the prototypes. This is the drop-top version of the car, with the soft-top neatly stowed. Fitting the heavy hard-top was really a two-man job.

Proving that the 2.5PI was not merely meant to be a sports saloon for drivers who insisted on changing their own gears, this one shows that automatic transmission was also available.

location for the vacuum servo unit (to clear the bulkier engine), while 185-13in Goodyear radial ply tyres were standardized. Add to this the Lucas 15AC alternator, twin-speed wipers and a leather-rimmed sprung steering wheel, and it was clear that the engineers and planners had done a thorough development job.

Only a saloon was available at first, though a 2.5PI estate car would be phased in, without fuss, in the spring of 1969. In general, prices had risen a little since 1963, but in October 1968 the 2000 range's line-up looked like this :

Model	Retail Price (including Purchase Tax)
2000 Saloon	£1,271
2000 Estate	£1,539
2.5PI Saloon	£1,450
overdrive (optional)	£ 65
automatic (optional)	£ 100

By any standards, the 2.5PI was an exciting car, for it was just as refined as the 2000 had

ever been, it was better equipped, and it had a 106mph (170km/h) top speed. Was it any wonder that *Autocar's* testers were impressed?

Perhaps the idea was obvious enough, but someone deserves a lot of credit for first authorizing the mating of a TR5 engine with a Triumph 2000 saloon. In the beginning an old hack demonstrator was converted, with practically a stock sports car engine dropped straight in place. It was meant to be a rally car, for a rally that never was (the RAC event of 1967 was cancelled because of foot-and-mouth disease) and *Autocar* tested it in January 1968.

To say that we were impressed is an understatement: we raved about it, like all those who tried it. Although we were sworn not to say so then, it seemed pretty certain that a production version would be announced . . . There is no doubt that this car has been designed for the enthusiast. It is the kind of car over which one is bound to be enthusiastic, and we are extremely pleased that it has been made by a British manufacturer and has turned out so well in so many ways. It is very satisfying to drive and to live with,

97

Triumph 2.5 PI (Mk I) (1968–9)

Produced
July 1968 to September 1969

Identification
Chassis numbers carried the prefix MD

Layout
Unit-construction body/chassis structure in steel. Five-seater, front engine/rear drive, sold as four-door saloon or five-door estate car

Engine
Type	Standard-Triumph six-cylinder
Block material	Cast iron
Head material	Cast iron
Cylinders	6 in line
Cooling	Water
Bore and stroke	74.7 x 95.0mm
Capacity	2,498cc
Main bearings	4
Valves	2 per cylinder, pushrod and rocker operation
Compression ratio	9.5:1
Induction/fuel supply	Lucas fuel injection
Max. power (net)	132bhp @ 5,500rpm
Max. torque	153lb/ft @ 2,000rpm

Transmission (Manual)
Clutch	Single dry plate, 8.5in diameter; diaphragm spring, hydraulically operated

Internal gearbox ratios
Top 1.00, 3rd 1.386, 2nd 2.100, 1st 3.28, reverse 3.369
Final drive   3.45:1
20.2mph/1,000rpm in direct top gear
Optional Laycock overdrive (on top and third gears) had a ratio of 0.82:1: overall ratio 2.83:1.
24.6mph/1,000rpm in overdrive top gear

Automatic transmission (optional)
Torque converter	Maximum torque multiplication 2.0:1

Internal Transmission ratios
Top 1.00, intermediate 1.45, low 2.39, reverse 2.09
Final drive   3.45:1
20.2mph/1,000rpm in direct top range

Suspension and steering
Front	Independent by coil springs, MacPherson struts, lower wishbones, telescopic dampers in struts
Rear	Independent by coil springs, semi-trailing wishbones, telescopic dampers

Steering	Rack and pinion
Tyres	185x13in radial ply
Wheels	Pressed steel disc, four-stud fixing
Rim width	4.5in

Brakes

Type	Disc brakes at front, drum brakes at rear, with vacuum servo assistance
Size	9.75in diameter front discs; 9 x 1.75in wide rear drums

Dimensions (in/mm)

Track	
Front	52/1,321
Rear	50.4/1,280
Wheelbase	106/2,692
Overall length	173.75/4,413
Overall width	65/1,651
Overall height	56/1,422
Unladen weight	(saloon) 2,632lb/1,194kg
	(estate) 2,744lb/1,244kg

When introduced in October 1968, the 2.5PI looked almost identical to the 2000, except that, from this angle, there was a circular 'PI' badge on the vinyl-coated rear quarter panel.

the kind you hate to park because it means you have to switch off and leave it.

Behind the scenes, though, Spen King was already perturbed by the quality problems his engineers were uncovering. In spite of their experience, Lucas could not produce a totally trouble-free installation, and this problem persisted.

The Lucas system was always troublesome, [King recently insisted] because I think Lucas didn't put sufficient weight of good engineers behind it. I think it could have been made better.

The worst Achilles heel was the pump itself, there was more trouble with the pump with the distributor unit. I remember being driven quite mad by the pump problem, and eventually saying that we couldn't go on, we simply had to ditch it because Lucas couldn't sort it.

I remember talking to Simms about pumps – we didn't get very far with Simms, maybe we should have done. Lucas was certainly complacent, their excuse being that it was all our fault – all the manufacturer's fault, because the buying department was only interested in price and not quality. I think both were guilty.

For a time, though, the public liked what they saw, and the original-shape 2.5PI sold extremely well. Although it was only officially on sale for a year (an estate version, barely advertised to potential customers, followed in February 1969), which meant that there was little time for word-of-mouth recommendations to get around, no fewer than 6,519 saloons and 223 estate cars were built. At one time during the year, one in three of all 2000-type cars being assembled at Canley were 2.5PI derivatives.

6 Rallying and Racing

I can remember, almost to the hour, when I decided to urge the development of the 2000 as a rally car. It was at the Novi control, during the Liège-Sofia-Liège rally of 1962, when I was running Standard-Triumph's motorsport efforts, and when three pristine works TR4s had started; on that day it looked as if none might finish. Why on earth, I remember thinking, were we using low-slung sports cars in such awful conditions?

Along with one exhausted mechanic, I had been slumped at the roadside in Novi Vinodolski, on the Adriatic coast of Yugoslavia for hours, hoping against all reasonable hope that all three TR4s would have survived the battering handed out by Bulgarian and Yugoslavian roads in the previous 24 hours.

It was a long and frustrating wait for almost everyone in the event, as only eighteen of the 100 starters would make it back to Belgium. A few yards away, the competent but (to me) almost ineffably smug Mercedes-Benz mechanics had serviced Eugen Bohringer's big 220SE saloon and sent it on its way to eventual victory. Ken James's works Rover P5 saloon, and no fewer than three works Citroen DS19s had all sailed serenely through.

Later – much later, it seemed – two of the TR4s arrived, looking rather sorry for themselves. There was time to give them attention, which they certainly needed. When they were jacked up, it was possible to squeeze underneath and see the appalling damage wreaked to their undersides by the rocks and ruts of the Balkans. The floorpans were pockmarked, the cooling louvres in the full length undershields had been battered flat, and on both cars the rear suspension spring pins were steadily forcing their way out of the side members. When we waved them off, I was not convinced that they would survive the next day's rallying. In the end, only one of the two cars made it to the finish.

From that moment I was convinced that the low-slung TRs were no longer suited for the latest type of rough, off-road rallies that works teams had to tackle. Even though I was already so tired that I had quite forgotten what day it was (and the drivers, let us be quite clear, had suffered a lot more than me), I realized there had to be a better way. I was jealous of the big saloons that seemed to make so light of such conditions: Triumph, no doubt, needed to match them with one of its own.

That was in August 1962, when the new Barb had already reached the prototype stage, but I knew of its existence. It was more than a year before the 2000 would be revealed, and exactly two years before it could ever tackle a rally. On that hot and dusty afternoon, though, it made its date with destiny.

LAYING THE PLANS

I had begun to run the works motorsport effort in February 1962, with Harry Webster in direct control, at first with TR4 sports cars, and latterly experimenting with Vitesse saloons. There had been some success, always on tarmac, and notably in the high mountains of France, Italy and Switzerland, but by the end of 1963 the works motorsport programme was at the crossroads.

The TR4s, though fast, stable and effective, were only suitable for tarmac rallies. Even then they were not quite fast enough to contend for outright victory; rough roads and loose-surface special stages did them no favours at all. The Vitesses had come and gone, too slow and too heavy to make much impression against the new British 'homologation specials'.

Harry Webster realized, even if his colleagues did not, that there was no future for either rally car at international level. For the second time in less than three years, therefore, he told me to sit down, think things through, and have a look at Triumph's motorsport future.

Harry was determined that Triumph should stay in top-level rallying, and he also wanted to take Triumph back to the Le Mans 24-Hour race, but before this he wanted to see all the alternatives spelt out. In view of the company's rather fragile profitability at the time, he also made me a quite startling promise: 'If I agree with what you recommend, I'll try to find the money to back it.' And if I recommended an ambitious programme? 'Don't worry about that. If it looks right, I'll get the company to finance it.'

What happened next has already been summarized in other books concerning the works Triumphs. Having pointed out three alternatives – to carry on rallying TR4s, to close down the department completely, or to start again with new models and redoubled effort – as an enthusiast I opted for the last option.

My recommendations reached Webster's desk in October/November 1963. I emphasized the way that international rallying had changed so much in only two or three years,

with two distinctly different types emerging. Some of the classic events remained, using all-tarmac sections, where ultra-special, high-performance machines could be expected to fight it out: for such events, I suggested, Spitfires developed in every possible way would be ideal.

More and more events, though, were taking to unmade roads and tracks, some of them very rough, and to meet those challenges Triumph needed to use big, solid, fast saloons of the type that Mercedes-Benz, Citroen, Volvo, Ford, even Rover and Humber, were already using. In the Triumph range, as I saw it, the brand new Triumph 2000 *could* be ideal.

It helped that Harry Webster was himself something of a rally enthusiast, and knew exactly where the trends were leading. To his eternal credit he read the paper, agreed with it, took it to his board, and came away with the backing for a much more ambitious programme for 1964. Not only did he back the idea of using the new 2000s in rallying, but he also agreed to see Spitfires in rallying *and* back at Le Mans for the famous 24-Hour race.

DEVELOPING THE CARS

First Steps

It was months before work could even begin on a 2000 motorsport development programme. First of all the road car had to be got into production and put on sale, then homologated; however, as only 1,000 needed to be built to get approval, and Standard-Triumph planned to make more than that before sales began in January 1964, this was not likely to be a problem. In any case the new Spitfires – race *and* rally cars – would take priority in my department for some time. The first true works entry for 2000s was not scheduled until the end of the summer of 1964, when the Spa–Sofia–Liège was to take place.

The 2000's debut in motorsport came in front of BBC TV cameras on 7 December 1963, when Colour Sergeant John Rhodes used this works car in the London Motor Club's Autopoint event. As can be seen, conditions were awful, but the 2000 was by far the best road car on this event, though outpaced by Land Rovers and similar off-road specials. It was an encouraging start to its competition career.

What you might describe as the very first works Triumph 2000 – though this was neither a seriously developed car, nor one which figured in a serious event – competed on BBC TV on 7 December 1963, weeks before regular deliveries of 2000s had started. The occasion was the London Motor Club's TV Autopoint, an event that combined off-road rallying with motorized orienteering!

Nowadays we would call it a 'made for TV' motoring event, for that is surely what it was. To suit TV requirements an event was set up on the loose surfaces of War Department land at Hungry Hill, near Aldershot, where London Motor Club members (including World F1 Champion Jim Clark!) competed against the British Army Motoring Association. A weird variety of cars were present, including trials specials, twin-engined Mini-Mokes, a Haflinger, Standard's old World War Two prototype Bug – and an early-production Triumph 2000.

In this event the object was for the vehicle to go from marker to marker as fast as possible, across the mud and the scrub, using tracks if and where they existed. For this event Triumph speedily prepared 2680KV, an absolutely standard vehicle except for undershielding for the engine bay and the rear suspension, and entrusted it to Colour Sergeant John Rhodes. It raced three times during the afternoon, winning twice and finishing second (to the twin-engined four-wheel-drive Moke on that occasion). It was no wonder that Colin Taylor, writing in *Autosport* after the event, thought: 'The Triumph 2000 of the Army team was the best saloon car and proved very competitive when it came to cross-country motoring.'

That, though, was no more than a happy diversion, which actually taught the works team very little. There was then a lull while other competition cars had to be prepared before work could start on the 2000s. Having reprepared three of the TR4s and shipped

them to North America to compete in the Shell 4000 rally, the competitions department then turned its attention to the Spitfires, for the Alpine rally and Le Mans. In fact it was not until June 1964 that the team could begin to think seriously about the 2000s.

Settling a Specification

Although the team only intended to use three cars at any one time, four brand new white machines with powder-blue roofs were collected from the production lines at Canley, so that one could be a spare, or a practice car. As was now traditional in this department, we reserved consecutive registration numbers for them: AHP 424B, 425B, 426B and 427B.

In the meantime, the team also scrounged the use of some old Barb development hacks from the experimental workshops next door to use as rally service cars. They might be second hand, but they would surely be a lot more suitable than ex-Monte Carlo Vitesses, under-powered Atlas vans, or Herald Estates!

One car, 5264VC (which carried the commission number X711) had started life in 1962 as the original 1.6-litre prototype, while 9081VC (X713) had been the second prototype. Even though these cars were definitely tired old machines, they were larger, faster and much more suitable as service cars than the motley selection of machines they replaced. Later, when 5264VC got too old even for this job, 5384KV (X720) took over its duties. This, please note, was before Triumph 2000 Estate cars were available, so in all cases, the rear seats were discarded and plywood boxes took their places, the better to carry mountains of spare parts.

Not only were they good service cars, but when chasing the Spitfires they were also useful for trying out bits and pieces for the forthcoming rally cars. Heavily laden service

cars soon gave us experience with springs and damper settings, though the special wheels and brakes were never tried out on these old machines. Surprisingly, they gave us no advance warning of the problems that were to strike the rally cars in the Spa–Sofia–Liège rally . . .

When the time came to prepare the rally cars, the intention was always to take advantage of current regulations and rallying trends by turning the works 2000s into rough-road 'tanks'. Near-standard cars, it was concluded, might be strong, but would not be competitive against the Lotus-Cortinas (which were lighter and a lot more powerful), while further-modified (Group 2) versions would be little better, but if the cars were always entered in Group 3 (the Grand Touring category where so many other much-modified saloons had found success), they might become formidable competitors.

When entered in Group 3 the rules allowed a great deal of change and improvement to be made. The basic architecture of the cars had to be retained, and the same basic body structure, engine, transmission and suspension systems had to stay in place, but almost any modification was authorized.

The biggest problem was not finding technical ways of solving the problems, but of convincing Sir Donald Stokes why it had to be done. It was not easy to persuade a tycoon that the new model of which he was very proud, was not going to be fast enough for rallying until it was modified almost out of recognition. There was more. Rallying enthusiasts could quite see why rowdy 2000s with Webers had to compete against Austin-Healeys and Porsches, but the general public could not. If and when the cars started winning, the company's advertising agents were going to have to work hard for their fees . . .

In the end, though, top management agreed to our proposals. Along with leading company engineers, therefore, Ray Henderson and I sat down to discuss (a) what was desirable, and (b) what was feasible. It was not necessarily the cost of what we proposed but the time it would take, that was the limiting factor.

In developing Group 3 cars, the object was

to make the 2000s considerably faster, stronger and with better brakes. We also wanted to make them impervious to awful road conditions. To hoist them even higher and further away from the damaging rocks of Yugoslavia and the British Forestry Commission stages, they also needed to have considerably more ground clearance.

Fortunately there were at least three engineers at Triumph – Ray Bates and David Eley (engine work) and George Jones (transmissions) – who saw this not only as a technical challenge but as a great deal of fun too, while John Lloyd, who had responsibility for the competitions department even though Harry Webster took a direct and controlling interest, always took paternal interest in what was brewing. In theory, in the design and experimental departments, all this specialized motorsport activity had to be achieved without disrupting any of the new-model programmes (the 1300, TR4A

and GT6 models were all on-going at this time). No one (not even Harry Webster) took a great deal of notice of such edicts, and with a great deal of goodwill the works 2000s were quickly transformed into thirsty but efficient 150bhp rally specials.

Well in advance of completion for their first event (Spa–Sofia–Liège in August/September 1964) the cars had been equipped with triple-Weber carb engines which pushed out 150bhp, *wide*-ratio gearboxes (with overdrive operating on top, third and second gears), limited slip differentials, 15in wheels and larger brakes – yet a lot more detail remained to be settled. For practical purposes, to cut down the glare on sunny events (and because it was fashionable and looked good as well) they were also given matt black bonnet and front wing crown areas.

Although Triumph had consulted Syd Hurrell of SAH Accessories when tuning the TR4s, the company did all its own engine

For 1964 and 1965 the works 2000s ran as Group 3 'Grand Touring' cars, with 2-litre engines fitted with three dual-choke Weber carburettors. These produced 150bhp, were effective and made a fabulous noise, even though fuel consumption was awful. At that stage the cast iron exhaust manifolds were retained, this being a homologation requirement.

development this time. Because the original twin-Zenith-Stromberg carburettor/manifold layout was impossibly restricting, an impressive array of three dual-choke Weber carburettors was chosen instead. These certainly liberated a lot of power (and who cared about fuel consumption, after all?), and memory suggests that they revved to at least 6,500rpm in that condition. Most of the torque increase was at the top end, when the cars were making a glorious noise – and service crews certainly never needed any other advance warning of their arrival.

With the works TR4s we had always been irritated by the fact that overdrive third and direct top gears were almost the same, and in the 2000 it was just as awkward – 4.66:1 compared with 4.1:1 on the standard car. For the Group 3 2000s, which used the same transmissions, we asked George Jones if this could be rectified.

Jones, whose curmudgeonly image was carefully cultivated and not at all genuine, was one of the brightest transmission specialists in the British motor industry. Almost at once he suggested that by asking his craftsmen to machine some new gears, he could arrange to provide a lowered third gear, which also made for a lowered overdrive third as well. When I thanked him for solving a difficult problem without fuss, he merely harrumphed, and growled: 'any bloody fool could have worked that out'! Any fool, perhaps, but *I* certainly hadn't thought of it . . .

Without overdrive being used this would have made the gearbox almost unusable, but when the drivers 'played tunes' with the overdrive switch it gave the heavy 2000 a totally practical seven-speed transmission. After that, we leaned heavily on Laycock's overdrive specialists to make sure that their complex box of tricks never let us down: it rarely did.

Although I knew that the cars needed more ground clearance, I was not willing to jack up the suspension to get the whole car higher off the ground; the rear suspension geometry would have been ruined if we had tried it. Instead, we specified 15in diameter TR4-type steel disc wheels, along with larger front-wheel disc brakes and special disc callipers. All this had the effect of raising the overall gearing, which we countered by using the lowest available rear axle ratio that Jones's engineers could machine – 4.55:1.

At the same time Harry Webster persuaded Salisbury Transmissions of Birmingham to provide American-type Powr-Lok limited-slip differentials, similar to those used on the works TR4s of 1963. Salisbury, no doubt, hoped that Triumph would eventually adopt these for production cars, though this was never done.

The finalized transmission was an excellent compromise – good ratios combined with standard casings and basic hardware, all very carefully assembled by the experimental transmissions department. Even so, there was always as much torque as this assembly could handle, and there were several breakages in the early years.

The result was a totally specialized and very low-geared rally car, which in spite of its great bulk accelerated like a moon rocket – steady at first, then ever faster. When *Motor* later tested a 150bhp works car – EHP78C – in 1965, it achieved the results shown in the table overleaf. On the rally car, the much wider gap between overdrive third and direct top gear was immediately noticeable. The use of overdrive second, too, filled in a mighty gap between second and third.

READY TO RALLY

Early Disappointment

When three cars started the Spa–Sofia–Liège marathon in August 1964, the works team had already carried out a lot of pre-event

	Works car	Standard 2000
0–30mph (50km/h) (sec)	3.4	4.1
0–60mph (100km/h) (sec)	10.4	13.6
0–80mph (125km/h) (sec)	18.8	26.8
0–100mph (160km/h) (sec)	33.9	–
Standing quarter mile (sec)	17.8	19.4
Top speed, overdrive top	111mph(179km/h)	98mph(158km/h)
direct top (6,500rpm on rally car)	105mph(169km/h)	94.5mph(152km/h)
overdrive third	91mph(146km/h)	89mph(143km/h)
third	75mph(120km/h)	75mph(120km/h)
overdrive second	61mph(98km/h)	–
second	50mph(80km/h)	50mph(80km/h)
first	32mph(51km/h)	31mph(50km/h)

testing, mainly using AHP424B. Everyone – especially the drivers – was confident. After 1962 and 1963, when the low-slung TR4s had battered themselves almost to destruction on the rocks of Yugoslavia, to use 2000s seemed to be an ideal solution. Naturally there was shielding under the engine and gearbox, but deflector shields were also fitted ahead of the rear semi-trailing suspension arms. Everything we knew about the Liege, and every bitter experience of the last two years, had gone into the preparation of these machines.

This, if only we had known it in advance, was to be the last, the fastest, and the toughest of all the Belgian-based marathons, for in the entire 96 hours there was to be only a single one-hour halt, at the turn-round time control in Sofia. Public opinion (in the shape of increasingly heavy holiday traffic) was about to kill off the event. In 1964, as ever, it was hot and dusty throughout, and 'road' (for which read unsurfaced tracks) conditions were rough. In its report, *Autosport* described the event as a Legendary Marathon – only 21 of the 106 starters finished the event.

For the moment, though, the works team faced up to the long event with great confi-

dence. My personal ambitions were to see at least two of the cars finishing, with at least one of them in the top five. The Austin-Healey 3000s were expected to set the pace, which the 2000s could not match, but there were high hopes of matching the works Rover 2000s and 3-litres, the Ford Cortina GTs and the Citroen DS19s.

Except that the Dunlop Duraband tyres that we were contracted to use seemed to be puncture-prone in Yugoslavian conditions, everything started well. Whereas most of the Rovers had disappeared before Sofia, the Triumphs were still going strongly by the time the route wound down through Titograd, on the Adriatic coast. At this point Terry Hunter's car was very well placed. Suddenly, between the Stolac and Split controls (with only a few hours of rough stuff remaining) all three cars broke down with the same problem – the rear suspension cross-beam mountings broke away from the floor pan pressings, deranging the rear suspension.

The cars could go no further, and it was no consolation to any of us that inspection showed this to have been a metal fatigue problem. When they eventually saw what had happened (it took days to patch up the cars and get them back to Coventry!), Triumph's engineers were aghast. Nothing like that had turned up in pavé testing at the prototype stage – so we were at least providing them with knowledge to make cars better in the future. The solution was to reinforce the floor pans in that area, this achieving complete success. It was the only time that a 2000 was ever forced out of a rally with such problems.

We had no excuses, and we made none. The fact that all the Rovers and Fords had failed as well, was little consolation – especially as three works Citroens made it back to Belgium.

The 1964 RAC Rally – Much Better

For the next few weeks the 2000s had to be ignored as the entire team then concentrated on the Spitfires, which produced fine performances in the Tour de France and Geneva rallies. For the RAC rally of November, however, all four works cars (at Harry Webster's request, AHP424B was loaned to Peter Bolton to drive) were prepared.

This time, too, there was an important innovation. For the very first time, Dunlop produced radial ply tyres with an off-road tread – SP44 Weathermasters – which gave Terry Hunter and me the opportunity to test them on the Bagshot loose-surface test track against tarmac-treaded SP3s.

Dunlop could only promise limited supplies (BMC, and Stuart Turner, were trying to hog all that they could for use on the works Austin-Healey 3000s), so in the end some difficult decisions had to be made. Using the SP Weathermasters, traction advantages on gravel were considerable, though tarmac handling suffered. Even so, since almost every one of the sixty-odd stages had loose surfaces we decided to use them, with excellent results.

This time, with their strengthened body shells, the 2000s were always competitive, though as Group 3 cars they would have to compete against the Austin-Healey 3000s. Even though their start numbers were spread from no. 35 to no. 65 (the Healeys started at nos. 11 and 14 – the seeding

Immaculate, and on the very first stage of the 1964 RAC rally, is one of the quartet of works 2000s. Note the raised suspension, the larger diameter (15in) wheels, the black anti-dazzle bonnet, and the roof-mounted spot-lamp.

process on the RAC rally was a bad joke in those days) their times were very encouraging. The rally started and finished in London. By the time the rally reached Oulton Park, after a day and a night, the 2000s were on the edge of the top ten, well ahead of the Rover 2000s, but not quite on a par with the Cortina GTs.

Jean-Jacques Thuner's car broke its rear axle gears in a Lake District special stage; this was a recurrent problem on these cars, which would persist for years, and no amount of detail development seemed to make the diffs bomb-proof. Bolton's loaned car was not being driven very fast, but Terry Hunter and Roy Fidler revelled in the wet and foggy conditions.

Breakfast, and a one-hour halt after two nights on the road was taken at the magnificent Turnberry hotel near Ayr, where an astonishing incident robbed Terry Hunter of fourth place in the final standings. Having clocked in and left the car in *parc ferme*, Terry breakfasted and fell asleep. Unhappily, when the time came for him to leave he was still asleep and was then penalized for a late departure!

No amount of lobbying (or an official protest, which I felt obliged to make) could make the officials change their minds. Hunter blamed me for not waking him up, I blamed his co-driver Patrick Lier for not being alert – but the fact was that we were all at fault. All of which left Fidler's car as the best-placed 2000 for the next three days.

After a monumental tour of Great Britain (these days, how many rallies include stages in the West Country, in the Highlands of Scotland *and* in East Anglia, all in five days, on one route?) the last stage of the event was held in the Rendlesham forest, close to the USAF Woodbridge airforce base east of Ipswich. On this stage Makinen's Healey, which led the 2000s in its class, suffered a split gearbox casing, and for a time it looked as if it would never stagger to the finish in Central London. Roy Fidler, who stood to gain most by its retirement, says that he navigated the last sixty miles by the smell of Castrol R – certainly Makinen's co-driver, Paul Easter poured gallons of oil into the box, which almost as quickly leaked it out on to the road. In the end Roy Fidler had to be satisfied with sixth place overall – the best Triumph showing on an RAC rally since 1958.

Success and Breakdowns

For 1965 the works team was strengthened by the arrival of Simo Lampinen, a young Finn who had already won the 1000 Lakes in a front-wheel-drive Saab, and once again the works effort was to be split three ways. After I moved out of Standard-Triumph, Ray Henderson and Gordon Birtwistle managed the team.

Simo Lampinen, the young Finnish driver, joined the works rally team for 1965, specifically to get his hands on the Spitfire, but he also enjoyed wrestling with the 2000s. This was the specially staged signing ceremony with Harry Webster in October 1964.

Originally it had been planned that Spitfires would be rallied on tarmac, while the 2000s would be rallied on endurance events, which was good, in theory, except

---

The works competition cars
Even in the 1960s and 1970s, quoting a registration number was not always a foolproof way of identifying a competition car. With that cautionary word, however, this is a complete list of the identities used by the works teams between 1964 and 1971:

2000 Mk 1

    AHP424B
    AHP425B
    AHP426B
    AHP427B
    FHP992C
    FHP993C
    FHP994C

All these cars were prepared and run from Coventry.

2.5 PI Mk I

    FHP993C (converted from 2000)
    GVC689D

These first two cars were Coventry-based. All later cars were prepared/maintained at Abingdon.
    UJB643G
    VBL195H
    VBL196H
    VBL197H

2.5 PI Mk II

    WRX902H
    XJB302H
    XJB303H
    XJB304H
    XJB305H
    KNW798 (Kenyan identity, Abingdon built)
All these cars were Abingdon-based.

---

that the Spa–Sofia–Liège marathon was cancelled! No money could be found for the team cars to tackle the Acropolis rally of Greece, and though Simo Lampinen wanted to tackle the Swedish and 1000 Lakes rallies, the budget would not stretch to that either.

The 2000s therefore found themselves tackling events where classes, handicaps, or both made them attractive. This explains why two cars started the Tulip rally (all-tarmac, hill climbs and sprints, and marking by a complex class-improvement formula), and why John Sprinzel drove a car in the Alpine rally, extra to the fleet of Spitfires.

Because the Weber-carburetted 2000s were always running in the Grand Touring (Group 3) category, this was always a risky strategy. In many cases there would be works Austin-Healey 3000s running in the same class – and these were truly formidable machines – while the GT category itself was well-populated with cars like the Alfa Romeo GTA.

On the Tulip rally, held in May, where handicaps counted for much, fortune smiled on the 2000s. Jean-Jacques Thuner's car not only won its class and finished third in the entire GT category (beaten only by the two works Hillman Imps), but finished ahead of the Morleys' Austin-Healey too. Later in the year, though, Sprinzel's Alpine rally car had to retire when the gearbox progressively lost all its intermediate gears.

Then came the RAC rally, where the 2000s were expected to do well. Not only were four cars entered, but as an experiment Fidler's car ran in the Group 2 category, which meant using Zenith-Stromberg carbs, standard-size wheels and brakes, and having a lot less power. Snow began to fall soon after the start, which made good traction an essential. This time, too, the seeding was better, for Simo Lampinen started from no. 6, the other cars running at 23, 51 and 54.

After the first 24 hours Lampinen's and Fidler's cars were well up among the leaders,

In a stirring performance in the 1965 Tulip rally, Jean-Jacques Thuner's 2000 not only won its capacity class, but finished third overall in the GT category. EHP78C was a newly built left-hand-drive car, the only such works machine assembled.

The 2000s were at their best in the loose, but with 150bhp they were also surprisingly nimble.
AHP424B on the 1965 RAC rally, ahead of a much lighter Cortina GT.

and by the second evening, after the Welsh sections, Lampinen was fifth overall. Following the passage of the Yorkshire forests (where there was truly thick snow), at a control near Middlesbrough Lampinen was challenging for the outright lead. Soon after this, though, his engine blew a headgasket, leaving Fidler's sister car to carry on the fight.

Two days later, having slid off the road several times (to quote *Autosport* : 'Someone suggested that the reason Triumph 2000s have four headlamps was that you could wipe off the outside two and still have two left'), Fidler's near-standard car finished magnificently in fifth place behind three front-wheel-drive cars and Makinen's Austin-Healey. Even more important was the fact that he had beaten the Rover 2000s and Cortina GTs.

Only two weeks later there was a flourish at the end of the season, which I actually witnessed from the co-driver's seat of Roger Clark's works Lotus-Cortina. The second Welsh international rally of the year saw a head-to-head fight between Clark and two of the 2000s driven by Lampinen and Fidler.

Right from the start of the event, no other cars looked like getting a look-in, with each of the crews setting outright fastest stage times from time to time. After the first night the Ford led narrowly, but Lampinen fought back strongly. If he had not gone off the road for several minutes in the fog in Llanbed forest, he would surely have won, but in the end the Triumphs finished second and third overall.

Back at home, Triumph also lent Roy Fidler one of the cars to tackle the British

In absolutely appalling conditions, Simo Lampinen's performance in the 1965 RAC rally was remarkable. At one point on the second day he led the entire event until forced out with engine failure. Body roll never stopped the young Finn from extracting everything from the big 2000.

Roy Fidler and Alan Taylor used AHP426B, reconverted to be nearer to standard Group 2 form, to win their class and finish fifth overall in the 1965 RAC Rally. Fidler complained about the car's lack of pace, but not about the result he achieved.

Jean-Jacques Thuner had already crumpled the near side of AHP424B when this Welsh stage was tackled in the 1965 RAC rally. He finished second in class, after which the car was sold off.

series – international *and* national-status events – most of the time with me as his co-driver. This was to be a frustrating season for all concerned, for although Fidler was extremely capable and the car was fast enough to win events, it was desperately unreliable, often with the dreaded broken axle disease.

On the first Welsh rally of the year (there were two that season), the car led for many hours until an ignition points failure let it down. Fourth place on the Shunpiker was followed by a win in the Welsh Marches, but a headgasket blew in the *Express & Star*, and Fidler put the car off the road on the Nutcracker.

Fidler then took second in class in the Circuit of Ireland, but the car broke its axle when on the way to the start of the Scottish (and never reached Glasgow!), and repeated the trick by breaking a half-shaft universal joint while leading the prestigious Gulf London event. By the autumn the stuffing had gone out of this programme, and only those fine drives in the RAC and the second Welsh raised his spirits again.

THE END OF THE GROUP 3 CARS

For 1966 the FIA introduced new Appendix J regulations. This meant that the works 2000s and Spitfires were badly hit, which made life very difficult for Ray Henderson. In international events saloons would no longer be able to compete, much modified, as Group 3 cars.

Board approval was given to a £55,000 competition programme (this was a much lower budget than in the previous two seasons), where 2000s would only compete in the Monte Carlo and RAC Rallies, and support would be given to entries in the East African Safari.

Donald Stokes was not at all happy about the success so far achieved (as far as he was concerned, unless a car was winning outright motorsport was a waste of money), stating that the results achieved in the USA (by 'Kas' Kastner's TR4s and TR4As) warranted the expenditure, but he was doubtful as to whether in this country the expense was warranted even if we were completely successful.

In the event, the idea of sending works cars to the Safari was abandoned, though £4,000 was made available to Leyland East Africa to back their own cars. It did not help – four locally prepared 2000s started the event, but none of them finished.

Because the rule changes obliged the 2000s to run with the Zenith-Stromberg carburettors and standard manifolds, they were rendered uncompetitive against the Lotus-Cortinas and other 'homologation specials'. As I have already noted, Harry Webster briefly considered putting a limited-production 2000 'hot rod' into production, but this never came to anything.

In any case, the works team's first event of 1966 – the Monte Carlo rally – also ran to its own very constricting rules, which required virtually-standard cars to be used.

Accordingly, Ray Henderson's mechanics had to build up a set of brand new cars, which had little more than 90bhp (instead of the 150bhp of the 1964–5 examples), standard gear ratios, and no limited-slip differential.

Although teams like BMC and Ford (with Mini-Cooper S and Lotus-Cortina types) were much better placed under this system, Triumph realized that they would be struggling. They *might* be able to beat the deadly rivals from Rover, but they had no illusions about their pace against the big Citroen DS21s and the Lotus-Cortinas.

It was not a happy event for the team, for although Roy Fidler finished fourteenth overall (and fourth in the unofficial 'front engine/rear drive' category), the two other cars retired, one with a broken gearbox, the other with a smashed differential. Even this result flattered Roy, for this was the year of the disqualification fiasco, when a number of leading cars (the works Minis among them) were ruled out after the event.

Even though the 2000s had not been involved in the scurrilous events, Sir Donald Stokes was not amused. As in later years, he saw little point in spending money for 'his' cars to finish out of the running, nor was he interested in sponsoring the sale of special cars to rectify that situation. Within weeks, therefore, the works effort was run right down. Cars were sold off, drivers were released from contracts, and the department effectively went into suspended hibernation.

Two programmes, however, survived – one for racing and one for rallying. Roy Fidler arranged to buy one of the cars for use in British rallies (FHP993C, his Monte Carlo car, with a Weber-carburetted engine, but running on 13in wheel suspension), while Bill Bradley (who had been racing a Triumph Spitfire in 1965) started a British Saloon Car Championship project with a fuel-injected car! Both had a measure of factory backing and both were prepared in Coventry. Fidler recalls that:

The handling was certainly very different on the 1966 car. The early Group 3 cars handled very well – they were great big tanks but they handled well. Perhaps it was because the later car was a lot heavier that they were harder to handle. Certainly it was OK in the dry, but as soon as it rained, well, on tarmac it was just as if I had great cobs of butter instead of tyres on the corners.

RAC Rally Champion

In those days the RAC Rally Championship was not a tight little series of international events, but a mixed bag including the RAC rally itself, the home Internationals, and lesser status events too. As the season progressed Fidler found his 2000 and himself progressively better placed, so an exciting end to the year was in prospect. His co-driver, Alan Taylor, had already spent time in the works team with Rob Slotemaker, and was a real road-rallying expert.

On the *Express & Star* event, the 2000 took second place, then three weeks later the same car finished fourth behind three BMC and Ford 'homologation specials'. Fidler did not tackle the Scottish, and on the incredibly fast and gruelling Gulf London event (I followed it, as a press man, in a Jaguar E-Type – and needed all the performance to keep up!) he was forced to retire when lying third behind works Mini-Cooper S and Saab models: a front suspension strut was badly bent on one stage, and later the engine failed as a result of the exhaust system becoming flattened.

In August he won the Bolton rally (held in mid-Wales), followed up with victory in the Cavendish, but then put the car off the road on the Rally of the Vales. Victory in the Bournemouth rally (held in Hampshire and Dorset) was only a prelude to the last two major events of the year – the RAC and Welsh Internationals.

Completely refurbished for the RAC, FHP993C had to run in Group 2 form, and was ludicrously seeded at no. 40. When the rear axle began to make awful grinding noises after only a few hours all seemed lost,

FHP993C started its career in the 1966 Monte Carlo rally, where Roy Fidler finished fourteenth overall. He then used this car, and AHP426B, throughout the season, winning the RAC Rally Championship of that year. The RAC MSA finally presented him with a trophy for that victory in 1994!

Roy Fidler (left) and Alan Taylor won the RAC Rally Championship in 1966 with FHP993C. The trio were reunited in 1994 . . .

but this was changed, and the car began creeping up the board. Then the troubles set in. An eight-minute 'off' in a Scottish forest was followed by headgasket failure (again!), and retirement.

For the Welsh rally, which followed only ten days later, the 2000 was given a fresh engine, reprepared, and sent to do battle with works Cortina-Lotus, BMC Mini-Cooper S and Rootes Rallye Imps. This, in fact, was the 2000's finest hour. Right from the start in Cardiff Fidler was on the pace, other works crews rapidly fell by the wayside, and after twelve hours he lay in third place. By the time the event turned south, from north Wales, the battle was between the 2000 and Tony Chappell's Lotus-Cortina. The 2000 set several fastest times, it led entering the Eppynt (tarmac) complex of stages, but eventually finished second overall – just 38 seconds adrift after thirty-two stages. To this day there is doubt about some of the winning car's stage times. It was being driven by a Welshman, in Wales . . .

With this performance Fidler and the Triumph 2000 took the RAC Rally Championship of 1966, though there was no

. . . by which time FHP993C was a 2.5PI prototype, as prepared for the 1967 RAC rally, and Roy boisterously admitted to being twenty-eight years older.

fuss. As Roy told me: 'I never got a letter of congratulation from the RAC, I never got anything. There was no recognition in any shape or form. The factory, though, was very good. I was pleasantly surprised.' Naturally Standard-Triumph, and Harry Webster, were delighted with the result. Webster asked Fidler to visit him at the factory ('would you like to come and have a word with me, we'll have lunch?'). All unsuspecting, Roy travelled to Fletchamstead North where: 'As I walked in, he had organized a press conference, where he presented me with a silver cigarette case, which had engraved on it: "Presented to Roy Fidler by the directors of Standard-Triumph International Ltd in recognition of his brilliant achievement in winning the 1966 RAC British Rally Championship in a Triumph 2000." By the way, there was a cheque for £100 in it!'

RACING IN 1966

British Saloon Car Championship racing

was always exciting and very popular with the crowds, but in the early 1960s when it was run to FIA Group 2 regulations the 2000s were not competitive. Winning cars included specially homologated Ford Lotus-Cortinas and American Fords – Galaxies and Mustangs – with enormous engines. For 1966, however, the series organizers, the BRSCC, bowed to pressure from other manufacturers and changed the rules to cater for Group 5 saloons instead. Although this still meant that standard cars had to be used as the basis, there was freedom for unlimited engine, gearbox, braking and suspension modifications.

Bill Bradley, who had successfully raced works-prepared Triumph Spitfires in 1965, immediately saw that a Group 5 2000 might be very fast, approached Harry Webster, and gained approval for a 'works'-backed programme in the 1966 series.

Using one of the 1966 ex-Monte Carlo rally cars, lowered, lightened and stiffened as much as possible, Henderson's team prepared the lightest, fastest and best-handling

One of the development 'tweaks' tested on 2000s, with competition in mind, was the fitment of inboard disc brakes at the rear. This picture dates from 1966, the car in question being Bill Bradley's race car.

Using a specially developed Lucas injected version of the 2-litre engine, Bill Bradley raced this works 2000 in the British Saloon Car Championship of 1966. Note the ultra-wide Minilite wheels and the lack of bumpers. The car looks nude because advertising and sponsorship was banned in those days.

2000 yet. The 2-litre engine, which was really a further-modified version of that intended for the GT6R Le Mans race project, which had just been cancelled, was given a new-type full-width cylinder head (a different, deeper-breathing casting, which would eventually appear on all these engines), an early example of the Lucas fuel injection system, and reputedly had at least 170bhp.

All the 1965-type Group 3 chassis and transmission equipment was used, along with new-type ultra-wide-rim Minilite magnesium alloy road wheels, but in spite of the effort put in there were still two major problems – one was that the car was still a lot too heavy, and the other was that it was not at all as specialized as the Lotus-Cortinas and the ex-Monte Falcons with which it was obliged to compete.

Throughout the nine-event season Bradley struggled to make the 2000 competitive, but although its engine always sounded gorgeous, it never had the straight line speed to match its rivals. Bradley's best performance was a third in class (behind Jacky Ickx and Paul Hawkins's works Lotus-Cortinas) at Crystal Palace in May, but the fuel injection and transmission gave a great deal of persistent trouble, which could never be sorted out during the season. Not even a change of driver could help – for the Motor Show 200 meeting at Brands Hatch in October, Roy Fidler was given a try, but the car retired with misfiring on that occasion. At the end of the year the project was quietly written off, this being Bradley's last link with Triumph in motorsport.

After the saloon car racing project had been abandoned, for 1967 Roy Fidler had to soldier on in British rallying with virtually no help. Somehow, without telling too many people about it, Ray Henderson managed to get him some material support, but there was no money available.

Fidler used FHP993C throughout the season (the car he had bought after the Monte

Viscount 'Kim' Mandeville drove this factory-prepared 2000 in the East African Safari rally of 1967, without success. In that event the cars had to run in near-standard Group 1 or Group 2 form, which made the big Triumph much less competitive.

Carlo rally of 1966), managing fifth place on the Circuit of Ireland, and sixth place on the Gulf London marathon (where he rolled the car – that was not easy, ask any other 2000 driver for confirmation! – and drove for many hours without a windscreen in place, with the crew wearing goggles).

By this time, in any case, progress had caught up with the 2000, for more and more special cars (like Ford Lotus-Cortinas) had been homologated. Except that a pair of excitingly specified TR5-engined prototypes were prepared for the RAC rally – and that story is told in Chapter 8 – the 2000 was near the end of its motorsport career.

The works 2000s, in International rallies

Year/ Month	Event	Crew	Result
1964			
August	Spa–Sofia–Liège	J-J. Thuner/J. Gretener	DNF
		R. Fidler/D. Grimshaw	DNF
		T. Hunter/P.Lier	DNF
November	RAC	P. Bolton/N. Baguley	Finished
		J-J. Thuner/J. Gretener	DNF
		R. Fidler/D. Grimshaw	6th overall, 2nd in class
		T. Hunter/P. Lier	3rd in class

1965

January	Welsh	R. Fidler/G. Robson	DNF
	Monte Carlo	P. Bolton/G. Shanley	DNF
April	Circuit of Ireland		
		R. Fidler/D. Barrow	2nd in class
May	Tulip	J-J. Thuner/J. Gretener	3rd overall GT, 1st in class
		R. Slotemaker/F. Geest	DNF
June	Gulf London	R. Fidler/G. Robson	DNF
July	Alpine	J. Sprinzel/W. Cave	DNF
November	RAC	S. Lampinen/J. Davenport	DNF
		J-J. Thuner/J. Gretener	2nd in class
		R. Fidler/A. Taylor	5th overall, 1st in class
		J. Sprinzel/D. Benson	3rd in class
December	Welsh	S. Lampinen/J. Davenport	2nd overall
		R. Fidler/A. Taylor	3rd overall

1966

January	Monte Carlo	S. Lampinen/J. Ahava	DNF
		R. Fidler/A. Taylor	14th overall
		J-J. Thuner/J. Gretener	DNF
	Sweden	R. Fidler/A. Taylor	DNF
	Circuit of Ireland	R. Fidler/A. Taylor	4th overall, 1st in class
	Gulf London	R. Fidler/A. Taylor	DNF
	RAC	R. Fidler/A. Taylor	DNF
	Welsh	R. Fidler/A. Taylor	2nd overall

Roy Fidler became RAC rally champion in 1966 with these and other performances in National-status events.

1967

April	Circuit of Ireland	R. Fidler/A. Krauklis	5th overall, 1st in class
June	Scottish	R. Fidler/D. Friswell	DNF
June	Gulf London	R. Fidler/B. Hughes	6th overall

7 The Mk II Models

History records that the restyled 2000 Mk II models were unveiled in October 1969, almost two years after British Leyland had been formed. Even so, it is a fact that the restyle had been proposed as early as 1966, and finalized in 1967.

When Spen King took over from Harry Webster, he found that the design had been frozen in all but a few decorative details. In those days, well before computer drafting had been invented, it took a long time to convert the shape of a clay model into the lines of a production car.

Even before work on a restyle began, Standard-Triumph knew that a completely new shell was financially out of the question (the company's body engineers, in fact, were probably quite relieved, as they were incred-ibly busy with other projects). Then, as later, the motor industry's 'bean counters' (the accountants) liked to get up to ten years of production, if not more, out of a single basic floor pan design. Whatever was agreed for a Mk II 2000 would have to evolve on the basis of the existing monocoque.

This was no bad thing, for first thoughts on the Barb theme had been totally correct (even though Harry Webster had always dis-agreed mildly with Michelotti in wanting a slightly wider car), and the design certainly did not need to be replaced. The original car looked good from every angle, the cabin was roomy enough (much more spacious, of course, than its major rival, the Rover 2000), and the basic running gear was adequate.

In any case, when the company got round

By the mid-1960s, Triumph – and Michelotti – were already thinking about a face-lift for the 2000 family. This was an early effort from Studio Michelotti, with the new and smoother front end already beginning to emerge.

to thinking about Mk II models, the engineering department – and Giovanni Michelotti – were already up to their eyebrows in new models, and at first there was simply not enough capacity to get the job done in a hurry. Michelotti's first proposals (sketches, not detailed drawings) were dated February 1967, but consider what was also happening in the design and development offices at that time :

Herald extended wheelbase versions were proposed.

Manx rear-wheel-drive version of 1300 being developed. This would eventually become the Toledo.

Vitesse new lower-wishbone rear suspension being considered.

Spitfire facelift/reskin proposal being worked up.

GT6 rear suspension, as Vitesse, being considered.

TR5 new six-cylinder-engined car, design work almost complete.

Stag first prototype in existence, engineering work going ahead.

Since the Leyland-Rover merger had just been proposed, efforts were also being made to shoehorn the new Rover V8 engine into various Triumphs, including the Stag. All this, by the way, was concentrated into a department so relatively small that in the 1990s every personality would probably be concerned with just one activity – meeting safety regulations!

DESIGN FACELIFT

During 1967 the engineers, planners, and sales force struggled to find an attractive balance for the Mk II model, which would eventually be coded Innsbruck. There was no doubt that what modern pundits call the 'platform' (the floor pan, the basic running

Component sharing

During the 1960s, Standard-Triumph engineers and planners (led by Harry Webster and George Turnbull) made sure that each basic building block (engines and transmissions in particular) could be used on several different types of car.

Although there were many detail specification differences for use with a particular model, this is how the 2000/2.5 model shared its building blocks with other Standard-Triumph cars.

1,998cc engine ** Standard Vanguard Six, Vitesse 2-litre, GT6 sports coupe
2,498cc engine TR5 and TR250 sports car, TR6 sports car
Four-speed all-synchromesh gearbox TR4, TR4A sports car, TR5, TR250, TR6 sports car, Stag, Dolomite Sprint, Spitfire (Le Mans car)
Rear axle/monocoque-mounted final drive TR4A, TR5, TR250, TR6 sports car, Stag, Spitfire (Le Mans car)
Engines and transmissions were also supplied to companies like TVR and Morgan for use in their own specialist cars.

** The smaller-bore 1,596cc engine was also used in the Vitesse 1600 range.

gear, the suspension, and the front bulkhead structure) would have to stay as it was, but then the bargaining began.

By this time Michelotti was being encouraged to develop a family look for all the new Triumphs likely to be introduced in the next few years. This, as we now know so well, was difficult to achieve at the front end of cars so diversely shaped as the Spitfire and the Dolomite, the TR6 and the 2000, but it was easier at the rear, where a whole series of early 1970s Triumphs were given sharply cut-off, vertical tail panels, with sharp lapped-over panels surrounding them. The Stag, designed in 1966, was the first to

Do not be confused by the 1964 number plate, which just happened to be lying around in Triumph's styling studio in 1967. This was just one of various treatments proposed for the new longer-nose Mk II cars.

Mk I versus Mk II shapes, side by side at a Triumph club meeting in 1994. Which do you prefer?

pioneer this total treatment, but because the launch of this GT car was delayed until 1970 the new family style actually appeared first on the Mk II 2000/2.5PI models, which had been designed afterwards.

To modernize the basic Barb shape without spending a fortune on press-tooling, the floor pan, wheelbase and the general layout of the suspension and wheels had to be left undisturbed. Work was then concentrated on making the revised car look different by changing the nose, the tail, and the facia and interior. It was agreed that all pressings for the main passenger cabin would have to remain unchanged, which meant that the doors, floor, roof, screen/front bulkhead, and rear window/rear quarter panels would all be carry over items.

Minor face-lifts of the front and rear ends were discarded, but because internal and structural panels had to remain unchanged, the only major new-shaped panels would have to blend in with the existing package. It was a difficult conundrum – as Rover was to find out when it, too, face-lifted its own 2000 in future years.

At Michelotti's suggestion both the nose and the tail were stretched – lengthened a little – for it would have been impractical to produce a different style if the same panel joints had been retained. When they viewed the finished product, experienced stylists could almost see how a completed car had had its nose and tail built up with clay to produce new aspects, for between the wheel-arches there was absolutely no change.

At the front of the car there were several attempts to find an appropriate theme before something very similar to the Stag was adopted. There was a different bonnet pressing (there was no central air intake on the revised car), and instead of a rounded nose over a wide-but-shallow grille, there was now to be a full-width grille. Headlamps and side/indicator lamps were new, the ensemble being nicely detailed.

The nose of the car was therefore stretched by 4.25in (108mm), while at the tail there was a similar lengthening – by 4.3in (109mm) – which incorporated horizontal Stag-type tail/side/indicator lamp clusters. The revised wing panels were swept smoothly back to the lipped-over edges. But only on the saloon: no changes, except in decoration, were ever made to the tail of the estate car shell. Incidentally, this increased the saloon's boot capacity by two cubic feet, but naturally there was no change for the estate car.

Overall, this made the new saloon 8.55in (217mm) longer than the old, but no more spacious in the cabin. In the 1990s this would probably be classified as a wasteful face-lift, but in 1969 no one complained. The difference in elegance between the Triumph and its obvious rival, the Rover 2000, became even more marked. Harry Webster, ever the pragmatist, presided over the update:

> There was always a limit to the capital available to us. On the 2000, for instance, changing outside skins was possible, but expensive tooling, things like door inners, window winding mechanisms, bulkheads, that sort of thing, had to be kept.

Spen King, Webster's replacement, who was not only an experienced engineer but a gifted stylist, has a different viewpoint today: 'I always liked the older-type 2000 better than the new. I thought that hanging bits of tin on the ends ruined the damned thing. I thought the old 2000 was a superb motor car.'

TOOLING BY KARMANN

By the autumn of 1967 the Mk II styles had been finalized, and at this point the company was hoping to introduce the new shape at the Geneva Motor Show in March 1969. Then, unexpectedly, they suddenly struck a big

Gordon Birtwistle must have been motoring seriously in these testing shots of a prototype 2000 Mark II, for he was wearing a crash helmet. From these angles, the amount of body roll under heavy cornering is obvious, as is the wider rear track.

This was the 2.5PI Mk II, as introduced in October 1969. Mechanically it was identical to the original 2.5PI, but shared the lengthened nose and tail of all other derivatives.

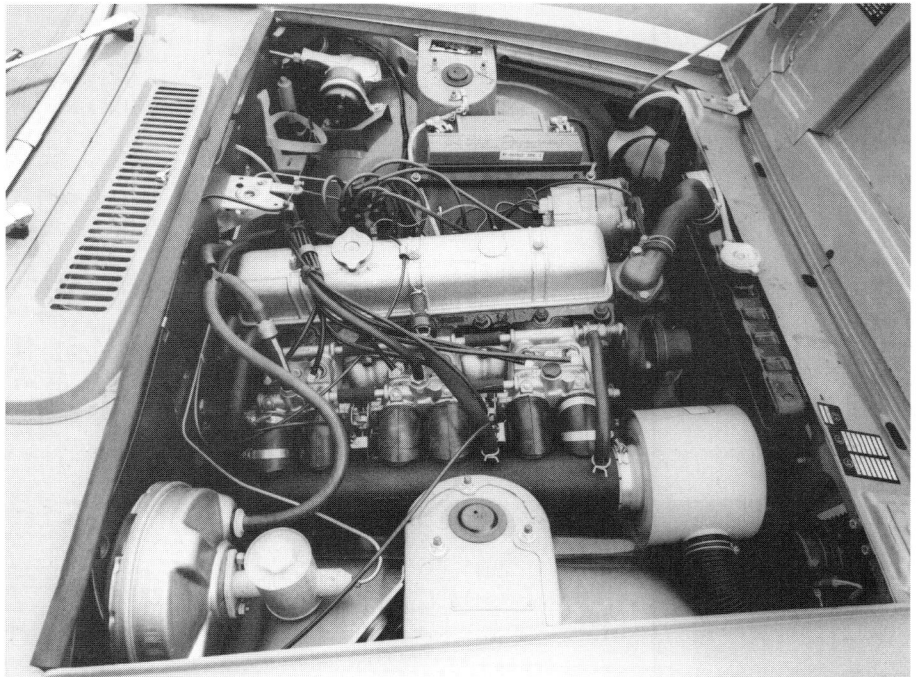

The engine installation of the facelifted 2.5PI was exactly the same as that of the 1968–9 variety. By this time the engine bay of these cars was beginning to look crowded.

snag. Needing to have new press tools and fixtures made, Triumph approached Pressed Steel Co. Ltd (who already produced body shells for the 2000) to tackle that job. To their utter consternation, Pressed Steel turned the job down.

Pressed Steel told Sir Donald Stokes that they could not possibly start at once, which meant that they could not do the job in time. However, they offered full co-operation if Triumph should make alternative arrangements.

Coincidentally, an alternative had already presented itself, and Pressed Steel's flexibility made it viable. The West German company Karmann, which had just agreed to produce new tooling for the TR6 model, was still actively looking for work in Britain. As the board minutes commented: 'As an alternative Karmann Ghia [sic] could supply our demands. They were short of work and their preliminary prices appeared reasonable . . .'

Karmann, therefore, soon found themselves contracted to provide the same service for the Innsbruck model as for the TR6. Karmann would make the press tools and any new jigs that were needed, and once they were ready these would be delivered to Pressed Steel at Swindon for manufacture of the revised car to begin.

But Karmann? Karmann who? In 1967 this company was not known outside Germany. What was special about this concern? Karmann, in fact, already had a fine reputation in Europe. By that time Karosseriefabrik Karmann was already one of the oldest coachbuilding firms in Europe. Founded as a bespoke coachbuilder in 1901, when only eight people were on the payroll, Karmann had converted to large-scale production after the Second World War, by starting to build VW Beetle Cabriolets in 1950, following up with the special VW Karmann-Ghia coupés and convertibles of the 1950s.

By the 1960s Karmann had won a BMW contract to produce 2000CS Coupé shells, and was already preparing to build the up-market 2800CS model when it won the Triumph contract. Not only were Karmann experienced, but they also worked fast – for the TR6 they would produce tools and jigs in little more than twelve months; for the Innsbruck model it would take longer, but the result would be just as satisfactory.

This, too, seems to be where the Innsbruck project code originated. Until Karmann became involved, company documents make no mention of *any* project code. None of the company executives I talked to could think of any other reason for the choice. As one said: 'Well, Karmann was German, and Innsbruck was Austrian. I suppose it was as good a secret code as any.'

UPGRADING THE SPECIFICATION

Even though the foundation of British Leyland threatened to throw Triumph's strategic planning into disarray, the Innsbruck project was never disturbed. Because of the lack of time, plans to launch the new model at the Geneva Motor Show had to be abandoned. Instead the new range was ready for launch at the Earls Court Motor Show in October 1969.

Development of the new model was so straightforward, in fact, that only two prototypes in the experimental X-series were ever built. There was, after all, no need to carry out much extensive pavé testing, and all mechanical work, testing and double-checking could be carried out on outwardly normal Mk I models. For the record, the two Mk II development cars – both saloons – were:

X785   Registered PVC918G. A left-hand-drive 2.5PI

Triumph 2000 (Mk II) (1969–77)
(renamed 2000TC from June 1975)

Produced
September 1969 to May 1977

Identification
Chassis numbers carried the prefix ME until 1974. Cars for Sweden carried the prefix MH. 2000TC models (1974–7) carried the prefix ML

Layout
Unit-construction body/chassis structure in steel. Five-seater, front engine/rear drive, sold as four-door saloon or five-door estate car

Engine
Type	Standard-Triumph six-cylinder
Block material	Cast iron
Head material	Cast iron
Cylinders	6 in line
Cooling	Water
Bore and stroke	74.7 x 76.0mm
Capacity	1,998cc
Main bearings	4
Valves	2 per cylinder, pushrod and rocker operation
Compression ratio	9.25:1 (from 1973 8.8:1)
Carburettors	2 Zenith-Stromberg 150CD or 150CDS (2 SU HS4 on later cars)
Max. power (DIN)	84bhp @ 5,000rpm (rerated 91bhp @ 4,750rpm from 1975)
Max. torque	100lb/ft @ 2,900rpm (110lb/ft @ 3,300rpm from 1975)

Transmission (Manual)
Clutch	Single dry plate, 8.5in diameter; diaphragm spring, hydraulically operated

Internal gearbox ratios
Top 1.00, 3rd 1.386, 2nd 2.100, 1st 3.28, reverse 3.369
Final drive 4.10:1
18.6mph/1,000rpm in direct top gear
Optional Laycock overdrive (on top and third gears) had a ratio of 0.82:1 : overall ratio 3.36:1 (0.797:1 with J-Type overdrive from November 1972, overall ratio 3.27:1).
23.2mph/1,000rpm (23.8mph/1,000rpm) in overdrive top gear
Final drive ratio raised to 3.70:1 (with overdrive, 2.95:1) from mid-1975, concurrent with fitment of 91bhp engine.

Automatic transmission (optional)
Torque converter	Maximum torque multiplication 2.0:1

Internal Transmission ratios
Top 1.00, intermediate 1.45, low 2.39, reverse 2.09
Final drive 3.70:1.
18.6mph/1,000rpm in direct top range

Suspension and steering

Front	Independent by coil springs, MacPherson struts, lower wishbones, telescopic dampers in struts; anti-roll bar from introduction of 2000TC in 1975
Rear	Independent by coil springs, semi-trailing wishbones, telescopic dampers
Steering	Rack and pinion (optional power assistance)
Tyres	6.50x13in cross ply (saloon), 175x13in radial ply (optional on saloon to 1973, standard thereafter; standard on estate car)
Wheels	Pressed steel disc, four-stud fixing
Rim width	5in

Brakes

Type	Disc brakes at front, drum brakes at rear, with vacuum servo assistance
Size	9.75in diameter front discs; 9 x 1.75in wide rear drums

Dimensions (in/mm)

Track	
Front	52.5/1,333
Rear	52.9/1,344
Wheelbase	106/2,692
Overall length	182.3/4,020 (saloon)
	177.25/3,908 (estate)
Overall width	65/1,651
Overall height	56/1,422
Unladen weight	(saloon) 2,620lb/1,188kg
	(estate) 2,750lb/1,247kg

X786 Registered PVC917G. A right-hand-drive 2000

Both were actually registered on 31 December 1968, less than a year before series production of Mk IIs would begin.

By that time, Triumph's production planners had done a magnificent job preparing for the change-over, for there was little recognizable hiatus in the flow between old and new types. The last Mk I-shape cars were assembled at Canley in September 1969, by which time the first few Mk IIs had already been built. To see how smoothly the change-over went, here are the month-on-month 'accounting period' production figures for the 2000/2.5 range, for the period June to November 1969, which are in any case distorted by the annual 'shut-down' in July/August :

Month	No. of Cars
June	1,430
July	1,547
August	774
September	848
October	1,491
November	1,573

When the press was introduced to the new cars, which were simply dubbed the Mk II models, much was made of the new styles and interiors for, it was emphasized, little needed to be done to the running gear of the cars.

Compared with the obsolete 1969-model

131

Facias and instrument panels of Mk II models compared. The 2000 had a neat style, rather devoid of instruments, with a two-spoke steering wheel. The 2.5PI had more instruments, including a rev counter, and an alloy-spoked sports steering wheel. The owner of the 2.5PI subsequently added yet another instrument (an oil pressure gauge) and a Kenlowe cooling fan to his car.

Mk Is, there were few mechanical changes. The engine of the 2.5PI was unchanged, while the 2000 engine was given the latest full-width cylinder head, along with an alternator. Gearboxes were unchanged, but the change-up points on the automatic transmission had been raised.

The major chassis improvement was the 2.5in (63mm) increase in the rear suspension track, this being done by providing new-type cast aluminium alloy semi-trailing arms, and longer drive shafts. At the same time, wheel-rim widths had been increased to 5in, there was a larger brake servo for the 2000, and a vacuum servo was specified for the 2.5PI derivative.

Even though company test drivers recommended it, there was still no front anti-roll bar in the specification. The suspension itself was softer than before, and those of us with experience of the Rover 2000 were convinced that Spen King's influence had been brought to bear in this aspect of the latest car. Not

only that, but power-assisted steering, by Alford & Alder, became optional. Like other mechanical improvements, this was directly related to work being done on the new Stag, though the public did not know anything about this model as yet, for its launch was still eight months away.

Cynics who knew all about the problems Triumph had experienced with 'stiction' in the drive shaft splines saw that nothing had been done to improve this on the new car. Indeed, it would not be until 1972 that the problem was even partly solved, when the factory specified a new type of molybdenum disulphide-based paste (Molykote G Rapid) to be smeared on the splines.

Fixtures and Fittings

Compared with the Mk I models, major improvements had been made to the interior of the car – to what estate agents selling houses might well call the 'fixtures and

Pausing briefly at John O'Groats on a Club Triumph endurance run in the 1970s, this 2000 has been further modified by its owner to include low-mounted driving lamps.

Like the original 2.5PI, the face-lift model retained the dummy-Rostyle wheel covers, the vinyl-covered rear quarter panels, and the 'PI' badges in the same location.

The 'PI' badge was always fitted to the rear quarters of fuel-injected big Triumphs.

fittings'. Although the basic dimensions of the passenger cabin and doors had not been changed, Arthur Ballard and Leslie Moore had given their design teams free rein to produce a new and (by 1969 standards) a very luxurious interior. It looked good, and, subtly but definitely, it also felt good to stroke and feel.

The centre-piece was a totally new facia/instrument panel layout, smoother, better equipped and altogether more nicely detailed than in any previous Triumph. Facia and door cappings were all decorated in smooth but unpolished wood veneer, and sets of new instruments included the circular multi-signal 'systems check' warning cluster that had already become familiar on the Triumph 1300.

Once again, Leslie Moore's team produced neat detailing for the interior of the second-generation 2000s and 2.5PIs – in this case including the ashtrays in the door trim panels.

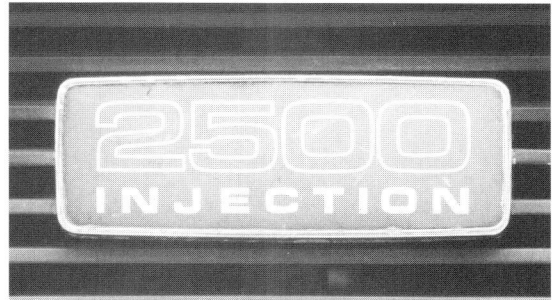

For the revised-shape 2.5PIs, Triumph were determined that everyone should know that fuel-injection was fitted. These are examples of the front and rear badges.

By the early 1970s, British Leyland roundels were ever-present on all their private cars, the Triumph 'world' having been banished some time ago.

Face-level ventilation eyeballs had been repositioned to the extremities of the facia panel, supplemented in the middle by a pair of fully adjustable rectangular outlets, while the centre console surrounded heater controls and the optional radio. The gear lever was given a leather gaiter, and where overdrive was specified selection was controlled by a neat sliding switch atop the gear lever knob itself.

On 2000 models the 120mph (225km/h) speedometer was balanced by a segmental three-in-one instrument showing water temperature, fuel contents and battery condition (this last replacing the ammeter fitted to older cars). On 2.5PI models the instrument display was more comprehensive, for the 140mph (225km/h) speedometer was matched by a rev counter, along with three separate circular instruments for water, fuel and battery condition.

There was more. The steering column had been provided with rake adjustment – the wheel could be moved up and down by no less than 4in (100mm) – there was new electrical switchgear on that column, and each version had its model type embossed into the padding on the steering wheel itself.

The seats, though retaining the same basic frames (as before, the fronts could be reclined), had redesigned covers with brushed nylon facings, a neater type of Britax static safety belts was specified, and inertia reel belts could be ordered as options.

Details like recessed, pull-out internal door handles, the rotary drum column control for lights and the automatic courtesy light for the opening boot lid all gave a good impression. Triumph, clearly, had looked closely at what the 2000 had already achieved, thought about it, shuffled their budgets, and came up with a functionally and visually more appealing machine. The overall effect was extremely pleasing, and was only achieved because of the experience of the body engineers, who were now

accustomed to working miracles within tiny development budgets.

If ever there was a case of a car that sold itself to the customer on the showroom floor, before he or she had even driven it out of the door, this was it, and Triumph dealers were delighted. Well before other companies caught on to this, Triumph and Rover were already adept at catching the eye; there was nothing new in this approach, however, as it had been honed on cars as early as the 1920s and 1930s. Spen King, new to the cars in 1968, was delighted to discover that:

Triumph had a very good bunch of people doing interiors. One of the snags with the SD1, which followed, was that at Rover and at Triumph they had old guys who did trim work, who had done a superb job, but somehow got kicked out of the way when SD1 was started.

Quite deliberately, I am sure, the marketing image of the new cars had been steered subtly up market, and this had to be paid for. Although UK prices, in general, had risen significantly since the original Mk I went on sale in 1963–4, the Mk II model prices still seem remarkably low by mid-1990s standards (*see* table below).

Model	Total Price (incl. taxes)
2000 Saloon	£1,412
2000 Estate	£1,686
2.5PI Saloon	£1,595
2.5PI Estate	£1,869
Popular options included:	
Overdrive transmission	£65
Automatic transmission	£101
Power-assisted steering	£52
Radial-ply tyres	£11

One advantage of the longer tail and more capacious boot of the Mk II models was that the spare wheel could be positioned under the floor, giving a lot more real luggage space on this later version.

The nose of the revised cars was altogether smoother than the originals, and was very similar indeed to the Stag, which appeared a few months later. Note the complete lack of a bonnet bulge on the revised cars. The front end looks smoother without bumper overriders, though owners who had to face crumpled noses after a shunt might not agree.

Compared with the originals, there were few engine bay changes on the revised carburetted 2000s. The alternator, replacing the dynamo, is one obvious improvement.

Only a small percentage of 2000s and PIs were ordered with the optional automatic transmission. This is a 1972 2000, with the quadrant control for the automatic in place of the normal 'stick shift'.

Enough, but not startlingly generous, rear seat leg room was provided in the revised models. Because the cabin style was not changed, there was no packaging improvement from one model to the other.

139

Triumph 2.5 PI (Mk II) (1969–75)

Produced
September 1969 to July 1975

Identification
Chassis numbers carried the prefix MG at first. 1974–5 models known as 2500PI (built alongside 2500TC models) carried the prefix MN

Layout
Unit-construction body/chassis structure in steel. Five-seater, front engine/rear drive, sold as four-door saloon or five-door estate car

Engine
Type	Standard-Triumph six-cylinder
Block material	Cast iron
Head material	Cast iron
Cylinders	6 in line
Cooling	Water
Bore and stroke	74.7 x 95.0mm
Capacity	2,498cc
Main bearings	4
Valves	2 per cylinder, pushrod and rocker operation
Compression ratio	9.5:1
Induction/fuel supply	Lucas fuel injection
Max. power (net)	132bhp @ 5,500rpm
Max. torque	153lb/ft @ 2,000rpm

Transmission (Manual)
Clutch	Single dry plate, 8.5in diameter; diaphragm spring, hydraulically operated

Internal gearbox ratios
Top 1.00, 3rd 1.386, 2nd 2.100, 1st 3.28, reverse 3.369
Final drive   3.45:1
20.2mph/1,000rpm in direct top gear
Optional Laycock overdrive (on top and third gears) had a ratio of 0.82:1: overall ratio 2.83:1 (0.797:1 from November 1972, overall ratio 2.75:1)
24.6mph/1,000rpm (25.3mph/1,000rpm) in overdrive top gear

Automatic transmission
(optional)
Torque converter   Maximum torque multiplication 2.0:1

Internal Transmission ratios
Top 1.00, intermediate 1.45, low 2.39, reverse 2.09
Final drive   3.45:1
20.2mph/1,000rpm in direct top range

Suspension and steering

Front	Independent by coil springs, MacPherson struts, lower wishbones, telescopic dampers in struts
Rear	Independent by coil springs, semi-trailing wishbones, telescopic dampers
Steering	Rack and pinion (optional power assistance)
Tyres	185x13in radial-ply
Wheels	Pressed steel disc, four-stud fixing
Rim width	5in

Brakes

Type	Disc brakes at front, drum brakes at rear, with vacuum servo assistance
Size	9.75in diameter front discs; 9 x 1.75in wide rear drums

Dimensions (in/mm)

Track	
Front	52.5/1,333
Rear	52.9/1,344
Wheelbase	106/2,692
Overall length	182.3/4,630 (saloon)
	177.25/4,502 (estate)
Overall width	65/1,651
Overall height	56/1,422
Unladen weight	(saloon) 2,760lb/1,252kg
	(estate) 2,873lb/1,303kg

ROAD TESTING RESULTS

Sales began as briskly as ever, helped along by the praise heaped on the cars by the most prominent motoring magazines. However, testers immediately realized that the Mk II cars were heavier than the Mk Is, that the suspension was perhaps even softer than before, and that none of the cars was quite as quick as the originals.

Perhaps that explains why *Autocar*, in assessing the 2000, found it a 'quiet and smooth car except when driven hard', and felt that they had to recommend that: 'Fast drivers who value performance would certainly be wise to go for the Triumph 2.5PI rather than the "soft" 2000 . . . Though very smooth, the engine does lack punch at low speeds – it has to be revved fairly hard for lively performance.'

On the other hand the team found the handling '. . . even more predictable than before, though the difference is slight. The independent rear suspension . . . gives very good absorption of small bumps and surface irregularities, but appreciably stronger damping is needed. There is a lot of body sway on corners and excessive bounce over undulations.' Theirs was a car that reached 96mph (154km/h) in overdrive top, and produced a 'typical' 24mpg (11.8l/100km) of fuel economy.

Motor, which tested the self-same car (RDU471H), seemed to be rather more impressed, calling the new car: 'Reborn – yet mature', opening its test with the words: 'Middle age is something that most of us don't like to admit to, but in compensation comes a maturity that we hope others recognize . . . After six years of adolescence [the 2000] has moved into middle age with an air of comfort and refinement that it never quite achieved before.' In summary: 'This comfort and the high standard of trim set the latest 2000 up as a highly placed contender in the luxury middle-weights. A first-class car.'

If Triumph's rivals were not perturbed by what they read about this car, they must certainly have winced a little when they read what was thought of the 132bhp 2.5PI Mk II. This, it seemed, offered an even more seductive combination than the original PI had done.

Autocar were more enthusiastic about the PI (the same car that *Motor* had already sampled a few weeks earlier), for which they recorded a 106mph (170km/h) top speed and typical fuel consumption of 23mpg (12.3l/100km), than they had been about the 2000, though the testers commented that the price was 'now rather high, but still very good value in its class'. However, 'if ever a car grew up overnight, the Triumph 2000 is a good example' was a more mature judgement which the factory appreciated. Moreover:

> Compared with the earlier PI (and the same goes for the plain 2000), the latest suspension is much softer and more yielding *almost* to the point of being sloppy. We emphasize the almost qualification because despite apprehensions about the ride on French roads, the PI took it well in its stride, never floating and only bottoming occasionally with three people on board and a good deal of luggage. We found it a very satisfying car, one which stands up well against continental competition for quality, refinement and efficiency as much as appearance.

When *Motor* came to try a 2.5PI Mk II, they discovered a car that was clearly in good shape, for the testers described it as a 'sumptuous family hot-rod with major refinements inside and out; smooth impressive performance; easier to handle with optional power steering but lacks precision handling;

excellent brakes, ventilation and fingertip controls.'

Motor measured a 110mph (177km/h) top speed in overdrive top gear, plus 108mph (174km/h) in direct top gear, and a level 100mph (160km/h) in overdrive third. All this, along with 0–60mph (100km/h) in 9.7 seconds, 0–100mph (160km/h) in 31.3 seconds, and an overall fuel consumption of 22.2mpg (12.7l/100km), made this a very desirable machine by 1969/1970 standards. No wonder they thought that this uplifted it from 'a comfortable but staid-looking car with hot-rod engine into a sumptuous well-planned family sports saloon which we look upon as a worthy cut-price BMW 2500'. Make no mistake, that is high praise.

When writing about the 2.5PI, in this case with automatic transmission fitted, *Autosport*'s John Bolster had no doubts at all, and was positively ecstatic. Calling it 'Britain's best medium-sized car', he opened his account with these stirring words: 'Wake up British Leyland! You are making a car, which, except for size, is the equal of anything built in this country, and you ought to tell the world!' There was more of the same, culminating in:

> Cruising . . . remarkably quietly . . . in the 70 to 90mph [110–145km/h] band, it can suddenly be catapulted up past 100mph [160km/h] as the fuel-injection engine takes hold, overtaking other cars with remarkable ease. Almost everything about it is so very right, but it teaches two lessons in particular. One of these is that, for a medium-sized luxury car, the six-cylinder engine is still far ahead of the 'four', and the other is that fuel injection is eventually bound to come, for all except the very cheapest cars.

THE CLIMAX OF A CAREER

Production – and sales – increased as never before. Even in December 1969, traditionally

Spen King took over from Harry Webster as technical director in 1968, presiding over the introduction of the Innsbruck range, the Stag and all later development of the theme.

never a month to break records, no fewer than 2,029 cars left the Canley assembly hall. By May 1970, when the usual spring buying fever was at its height, this rose to no less than 2,454 cars. Triumph's (and British Leyland's) accountants beamed, the profits poured in to the company, and Triumph's sales force relaxed a little.

In 1970 a total of 32,074 Innsbruck cars of all types were produced – no fewer than 10,090 of them with fuel-injected engines – while 31,692 (10,697 being fuel injected) followed in 1971, but this was really the high point of the car's long career. More than 100,000 cars would be built before 1977, but the last cars were technically still very close to those of 1969–70.

The fuel-injected 2.5PI, in particular, was a continuing source of worry, especially to Triumph's service engineers, who had to

keep owners happy. Reliability problems with the injection system were never truly solved by Lucas (though by the 1980s smaller specialists seemed to have found ways around every problem), warranty claims were high, and on occasions there were engine bay fires that could cause terminal damage to a car. One of my friends, a Coventry-based solicitor, had his 2.5PI catch fire in the middle of the city, and after all attempts to douse the flames had failed, he abandoned the car to its fate and went home by other means. His profession was enormously helpful when dealing with the insurance claim that followed . . .

Although tentative work on a successor to the Innsbruck – the Puma project – had begun by this time, there seemed to be no impetus to turn good ideas into the metal, or even to make full-size clay models for styling viewing. Product planner Alan Edis confirms that little new design or development work was ever carried out on the 2000/2.5PI models after 1971, for effectively their evolution had already been put into the 'care and maintenance' category.

Triumph, in fact, was far more committed to developing a new model to replace Innsbruck than to continuing to make any more significant changes. However, this idea was soon to be frustrated. As I make clear in Chapter 10, during 1971 British Leyland had already decided to merge Triumph with Rover, this corporate move being formalized early in 1972. Almost at once the Rover-inspired SD1 model was cho-

sen to replace *both* of the competing Triumph and Rover models, and Puma was killed off.

Time, though, was still on Triumph's, and the 2000's, side. Getting the new Rover SD1 project from 'good idea' to 'production car' level was going to take at least four years, so there was still time for the Innsbruck range to be rounded out and finalized. Product planning studies, which had been an undeveloped art at Triumph until the 1970s, showed that there was still a significant gap – both in performance and in pricing – between the 2000 and the 2.5PI, and that it ought to be filled. The result, unveiled in 1974, was the 2500TC.

8 The 2.5PI in Motorsport

Between 1967 and 1972 the works 2.5PI rally cars made many headlines, at world level. For Triumph – and later for British Leyland – there were many opportunities to beat the drum. The cars competed all round the world, in many different types of event.

Not only did the cars appear in events like the *Daily Mirror* London–Mexico rally, the Safari, and the RAC rally, but great personalities like Monte Carlo rally winner Paddy Hopkirk, London–Sydney marathon winner Andrew Cowan, and Formula 1 World Champion Denny Hulme all lined up to drive the cars.

THE 1967 RAC RALLY

Their rallying career began more than a year before the 2.5PI was even ready to go on sale. Ray Henderson, who was running Standard-Triumph's very restricted motorsport programme at the time, saw that regulations for the RAC rally of 1967 included a category for Group 6 'Sports Prototypes', and realized that the 2000s could be made competitive once again.

The new fuel-injected six-cylinder TR5 sports car was almost ready to be introduced, the first prototypes of what would soon become the 2.5PI saloon were already on the road, so Ray speedily gained permission to prepare what were, in effect, TR5-engined 2000s for the event.

When Triumph suddenly revealed that not one but two 2.5-litre fuel-injected 2000 'prototypes' would start in the 1967 RAC rally – one of them to be driven by the newly crowned F1 World Champion, Denny Hulme – every Triumph enthusiast was surprised and delighted. Since Appendix J rules had changed at the end of 1965, effectively outlawing the use of the very special breed of works 2000s, the big Triumph saloon had no longer been competitive. Things had been very quiet at Fletch North since the 1966 Monte Carlo rally.

The decision to prepare these cars came about by chance, for there was certainly no strategy behind Triumph's motorsport efforts at this time. As Ray Henderson has told me on so many occasions, his tiny department was a hand-to-mouth operation at the time, sometimes working on competition cars, sometimes working on 'overspill' jobs for the development department – and sometimes wondering how to keep themselves busy!

Triumph's problem, at the time, was that the world of rallying and the regulations covering the sport had changed so rapidly that they no longer had a competitive car in the range. To win, or even to be competitive, specially developed cars that we had already nicknamed 'homologation specials' would have to be used. Throughout the 1960s I (running Triumph motorsport from 1962–5) and Ray Henderson (1965–9) had both made these trends very clear to Leyland and Triumph management, but investment in special new cars was never made.

Even though Triumph's supremo, Sir

Donald Stokes, was not a fan of spending money on rallying unless success could virtually be guaranteed, somehow, from somewhere, Harry Webster diverted enough money for Henderson's mechanics to prepare not one but two fuel-injected cars. One car, for Roy Fidler/Alan Taylor, was actually a rebuild of the well-used rally car (FHP993C) with which Fidler had already won the RAC Rally Championship, while the second, for Denny Hulme, was GVC689D, a 1966 ex-Press demonstration 2000 newly prepared for this job.

Packaging the modified, more powerful, TR5 engine into the 2000's engine bay was easy enough, and the rest of the technical specification almost wrote itself – it was a simple matter of updating the 1965 'Group 3' specification – but other significant detail improvements were also made. 5.5 x 13in Minilite wheels were available, though the usual 4.5in rim x 15in wheels were set to be used on rough road special stages. There was also a tyre supply problem to be resolved – Triumph (and Roy Fidler) always used Dunlop, yet Denny Hulme was contracted to Goodyear!

Because the cars were to run as Group 6 prototypes, front and rear bumpers were removed, cooling slots were added to the bonnet panels, the rear seat was removed, and the spare wheels (two of them) were stowed in pouches immediately behind the front seats. The steering wheel was non-standard, and there was a small binnacle on the crash roll that contained an electronic rev counter and engine oil pressure gauge.

The Recruitment of Denny Hulme

Even in those days, when racing drivers were not paid a king's ransom, Triumph could never have afforded to hire Denny Hulme without outside help. For this the company had to thank Barry Gill, motoring correspondent of *The Sun* newspaper, which was sponsoring the event. (*The Sun* also helped to persuade Graham Hill to tackle the rally, in a works Ford Lotus-Cortina.) Although I had left Standard-Triumph by this time, I was delighted to be asked to co-drive for Denny even though I had never even met him at that juncture.

I soon discovered that hiring Denny might be one thing, but actually pinning him down to a programme, and teaching him how to drive a rally car on loose surfaces was quite another. To quote his journalist friend Eoin Young, from an *Autocar* column of the period:

> Denny Hulme scorns the use of a manager, and consequently he sometimes gets caught out with a pile of paperwork problems. Take the three cars he was going to drive in the RAC rally as an example! At various stages Denny was listed to drive a works car for BMC, a works car for Ford, and when it was all sorted out he was driving a works-entered Triumph 2000 fitted with the 2.5-litre six-cylinder fuel-injected engine from the TR5.

Although the deal was done early in October (the two of us meeting in Ray Henderson's workshop immediately after Denny had visited the Jaguar factory to collect a new car), we all had to wait until Denny had clinched his new title in Mexico at the end of the month before he could get his hands on a car.

In the end Henderson cobbled together a 2000 development hack (GKV 305D), suitably braced with undershielding, and persuaded Denny (and me) to drive it round and round the loose-surfaced Alpine course at Bagshot to get the feel of it. This also happened to be the same day that Ford's Roger Clark and David Seigle-Morris were showing racing driver Graham Hill how to drive a Lotus-Cortina. Eoin Young now takes up the story about Hulme:

> He's been practising hard on the army testing ground at Bagshot in Surrey, flinging a

World motor racing champion Denny Hulme planned to do the RAC rally of 1967 in a Triumph 2.5PI prototype. The very first time he drove a 2000 of any sort was on a foggy morning in Richmond Park, followed by a rough testing session at the Bagshot testing grounds. Fast, but not sensationally so, he was looking forward to the event, which was cancelled at 12 hours' notice because of an outbreak of foot and mouth disease in the countryside.

'hack' 2000 round the forest tracks, being taught the G-defying (G for Gravity, not God) sideways antics of rallymen by people like Roger Clark. The Triumph slithered off into a bank after one unscheduled take-off, but Graham Hill went one better the same day, flipping a works Cortina . . . Denny has filed this incident for future reference, and as he checked out the comprehensive array of tools and spares of the rally Triumph he announced loudly that he also required 'a shovel and a bloody great hammer' . . .

Except that he hated the tumult of publicity stirred up by *The Sun* Denny faced up to the rally as professionally as expected. His main worries, he confided in me, was that he would be very slow at first (he did not actually drive the fuel-injected rally car until he was introduced to it at Scrutineering at London Airport's Excelsior hotel), that he would not be as fast as team mate/rival Roy Fidler, and that he would not be able to beat Graham Hill. Because the RAC's seeding system was virtually non-existent, Fidler was given no. 60, but since Denny was due to start at no. 12 and Hill at no. 9, there would be plenty of time for the F1 drivers to compare notes.

Last-Minute Cancellation

A look down the entry list showed that this promised to be an exciting contest, for the prototype category not only included the 2.5PIs, but Graham Hill's Lotus-Cortina, an alloy-blocked Austin-Healey 3000 for Rauno Aaltonen and a Porsche 911R for Vic Elford. In addition, special versions of otherwise familiar works cars such as the BMC Mini-Cooper Ss and Lancia Fulvia HFs were also appearing in this category.

On paper this was due to be a demanding exercise, for there were to have been sixty-nine special stages (473 miles/761km) in five days, with a route stretching from London to Dartmoor, all the way up to southern Scotland, and back to London – with no more than one overnight halt, in Blackpool. On paper, too, the 2.5PI-engined 2000s were probably as fast, in a straight line, as any other saloon car in the event.

As everyone knows, the event was cancelled at twelve hours' notice, when a spreading epidemic of foot-and-mouth disease among cloven-hoofed animals made it politically impossible for the event to be run at all. Even before then, wholesale re-routings had been proposed, cutting out several of the worst areas, which made Alan Taylor and me spend every spare minute working out alternative service plans. Denny, bemused by such mysteries, went home to Surrey to sit it out, and almost sounded relieved when I called him to say it was all off.

Although the organizers then put on a single-stage TV spectacular for the following day, Denny did not take part, and even though I actually talked to him at Indianapolis in mid-1968 about running again later that year, nothing came of it. Neither he, nor those cars, competed again in international rallies.

Triumph enthusiasts wanting to measure the potential of these cars should try to find archive copies of *Autocar*, dated 25 January 1968, where Michael Scarlett spread himself over three pages in testing the Hulmemobile. Considering that this was a heavy (2,900lb/1,315kg) car, the performance figures were remarkable, and much more impressive than *Motor*'s test of a Weber-carburetted 1965 car had been.

Top speed, really immaterial on a rally car, was 117mph (188km/h). Using a 4.55:1 axle ratio, complete with Powr-Lok limited-slip differential, this car sprinted to 60mph (100km/h) in 9.2 seconds, to 80mph (125km/h) in 15.1 seconds, and to 100mph (160km/h) in 24.5 seconds. The 0-80mph (0–125km/h) figure, for instance, was only half that recorded for the standard (90bhp) car tested in 1964. Even the fuel consumption – 17.9mpg (15.8l/100km) for 546 miles (878km) – was better than Ray Henderson had feared.

The cancellation of the RAC rally was one of the great disappointments in my life, but for neither of these cars to be used in rallying at the time must have been immensely frustrating for Ray Henderson and Harry Webster. It was, indeed, a cruel blow in more ways than one, for Sir Donald Stokes seemed to take the cancellation as yet another example of the unbusinesslike way that motorsport was run, thought the whole escapade was a waste of money, and closed down the Coventry competitions department completely – again! Until 1969, therefore, no further work on works 2000s or 2.5PIs was officially carried out (though the fascinating one-off 1300 4x4 rallycross car of the period featured a works 2000 rear suspension layout under the skin).

UNDER NEW MANAGEMENT

After Leyland merged with British Motor Holdings in January 1968 to found the British Leyland Motor Corporation, there was a period of great uncertainty throughout the new combine, at all levels and in all activities. In particular, no one seemed quite sure about what should be done in motorsport, and once Harry Webster had been moved to Austin-Morris at Longbridge, Triumph no longer had a motorsport nut in a position of power.

Donald Stokes was by no means a motorsport enthusiast himself. However, he seemed to accept that there should be *some* form of corporate motorsport programme, and decided to collect all that activity under one roof. Because BMC's operation was currently the largest and the most successful in

the new Group, the Abingdon workshops were chosen as the base. BMC's then competitions manager, Peter Browning, became British Leyland's motorsport chief. Of that time he recalls:

My brief was to liaise with all the chief engineers. It was all very formal – there was a written brief, a Policy Instruction with a certain number. Harry Webster moved over from Triumph to Austin-Morris at Longbridge. I was therefore dealing with Spen King at Triumph, Peter Wilks at Rover, and Harry Webster at Austin-Morris. The only two people who really had competitions experience were Ray Henderson, and Ralph Nash at Rover . . . I never actually got central control. I got an instruction that we had to look at using other cars in the group because we – BMC, that is – were not winning any more, the Mini was getting long in the tooth . . . we were told to shop around. I was given a super brief, which was to go to all these guys and find out what was in the cupboard, to find out what was coming in the future, and to try to evaluate a programme of competitions.

The problem was that they were only prepared to give approval for projects on a very short-term basis . . .

There was certainly pressure that Triumph was the name they wanted to push. It was a clash of philosophies . . .

In 1968 and early 1969 Abingdon was completely occupied with rally programmes for the Mini Cooper S and the BMC 1800 (where the Safari outing was a disaster, but the London–Sydney entry almost a total success), followed by a racing saloon car programme for the Minis. Then, after an assessment period, came the chance to use Triumphs again.

First of all Ray Henderson was sent out to observe the East African Safari, where two privately-prepared 2.5PIs had been entered by the local importers, after which:

Tired but happy – Viscount 'Kim' Mandeville (in check shirt) and S. Allison pose on their privately prepared and very travel-stained 2000 at the end of the 1968 Safari. There were only seven finishers – it was truly an epic event in which they finished third overall.

We dismissed other projects pretty quickly. Apart from the Range Rover [for intercontinental marathons], we considered the Rover 3500S for rallying. I felt that the Rover could be used for rallying, on the basis that there was no substitute for horsepower. Other Triumphs? Well, we were shown various new projects by Spen King including the Stag, God help us, which wasn't suitable.

After a great deal of testing, which Brian Culcheth carried out, it soon became clear that Triumph's new 2.5PI saloon was the most promising of the bunch, and this was chosen for development. Culcheth's memory is that 'it was the only suitable British Leyland vehicle we could see that had a suitable pedigree'.

The initial problem was that Ray Henderson had to stay in Coventry, and the cars would have to be built, developed, and run from Abingdon. The famous BMC mechanics had no Triumph expertise at all at that point and – equally importantly – no spares either.

To get their rally programme under way, the Abingdon team started using a car to what Peter Browning now admits was a 'Ray Henderson' specification. 'Ray Henderson was always very good,' Browning remembers, 'there were no chips on his shoulder, and I always found Spen King to be courteous, very friendly and very helpful.'

As it happened, the first works 2.5PI to start a rally was Coventry-built, but Abingdon-run and registered (UJB643G). Paddy Hopkirk/Tony Nash drove it on the Austrian Alpine of May 1969. This was not an inspiring start to the car's career, for the transmission gave trouble right from the start, and while Hannu Mikkola's works Escort Twin Cam drew rapidly away, the 2.5PI could never get into the top ten. After taking a couple of sixth fastest stage times near the end of the event, the car finally expired with a deranged clutch.

A few weeks later a new Abingdon-built Group 6 prototype car (which carried the same identity – such things have been rife in

This was the very first outing by an Abingdon-entered Triumph 2.5PI. Paddy Hopkirk and Tony Nash struggled with an unfamiliar machine throughout the Austrian Alpine rally, before it finally retired with a failed transmission.

Paddy Hopkirk (at the wheel) and Tony Nash in the first Abingdon-prepared works 2.5PI during the Austrian Alpine rally of 1969. It would all be valuable experience, paving the way to the building of cars for the Daily Mirror World Cup *rally of 1970.*

151

works teams for many years!) competed in the Scottish rally, where Brian Culcheth used the 165bhp machine to put up an altogether more startling show. Holding down third place for many hours (behind Simo Lampinen's winning Saab, and Chris Sclater's ex-factory Lotus-Cortina), then fifth after Culcheth put the car off the track in the Loch Ard stage, the big Triumph eventually clawed back to third once again.

Finally, on the fourth day (of five) the 2.5PI's rear axle failed, and the mechanics took ages to change a unit never intended to be changed (to quote *Motor*'s report: 'The removal of the old differential took just over one hour, and it was an hour and fifty-nine minutes before the car was driveable again') – the brackets had actually been *welded* to the trailing arm carriers – which meant that the car slipped back to twenty-fourth place at the finish.

Considering the short time the Abingdon team had had to get used to a new model, this was a very promising start, which encouraged everyone. Culcheth's post-rally report has survived: among the comments were that the engine power was good, and that the car could cope with more. Culcheth liked the handling, but was not impressed by the longevity of the dampers. He also commented that the rear window had started to come away from the shell (we would hear a lot more about that, in future years).

THE 1970 *DAILY MIRROR* WORLD CUP RALLY

News came of a prestigious event for which a rugged 2.5PI rally car might have been specially designed – the *Daily Mirror* World Cup rally of 1970. Following up the London–Sydney marathon of 1968, a further intercontinental marathon had been rumoured for some months before creator John Sprinzel found his sponsor – the *Daily Mirror*. Just a week before the 2.5PI's excellent showing in Scotland, the *Daily Mirror* revealed that it was to sponsor a 16,000 mile (25,000km) marathon from London to Mexico City, to coincide with the build-up to the World Cup football competition.

Peter Browning had been clever. To help Sprinzel with his route surveys, he had loaned the ex-Scottish rally 2.5PI, on the understanding that it would come back to Abingdon with a report on its behaviour (Ford's reaction to this coup is not recorded). This car was later used in rallycross, and also for more World Cup testing.

Preparation

Almost at once Peter Browning then began the search for funds to support a full British Leyland entry in this potentially phenomenal event, and once Ford confirmed that it would also enter the event this was approved. As Browning confirmed a few years later, preparation for this Marathon was *the* single most important project to be considered at Abingdon for the next year. 'We had a very good budget for the event, and Lord Stokes was excited about it,' Browning records. 'We had to say – we did say – that we could win. We wouldn't have been allowed to go at all if we were not sure to be front runners.'

By the autumn of 1969 Browning's team was totally focused on the World Cup rally, and to get some experience under their belts three Mk I-shape 2.5PIs were entered in the RAC rally. These ran in Group 2 form, with about 140bhp. On an RAC rally with normal late autumn weather at least one of the cars might have finished in the top six, but snow fell heavily, particularly in Scotland and Wales. All three cars made it to the end, taking the first three places in their class, but the most successful driver, Andrew Cowan, could only take eleventh place overall.

Team-mate Paddy Hopkirk did not enjoy

The works 2.5PI Mk Is were rallied only once – in the snowy RAC rally of November 1969. Driving VBL195H, Andrew Cowan finished eleventh overall and won his class, while Paddy Hopkirk wrestled with the lack of traction in VBL197H to follow Cowan home.

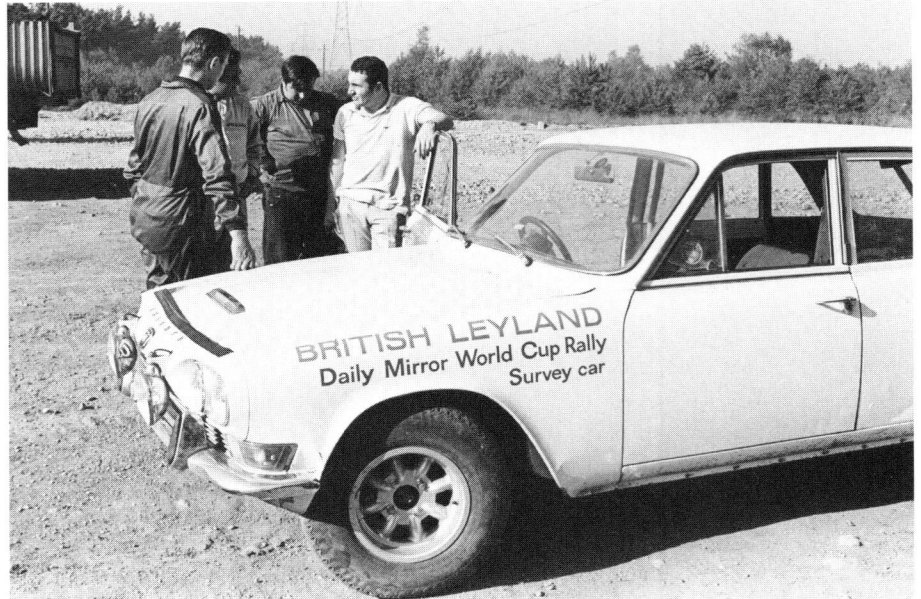

After the RAC rally of 1969, the Mk I-shape 2.5PIs were used as survey and practice cars for the Daily Mirror World Cup *rally. They were sent to South America for many months. Here Paddy Hopkirk (right) and Brian Culcheth (centre) discuss the cars' specification with two of the mechanics from the Abingdon competitions department.*

himself, and to a TV interviewer who wanted to know what was the hardest thing about the rally, he quipped, 'Winning it in this motorcar!' Amazingly, he was neither sacked nor disciplined, for he was very definitely Abingdon's favourite son at that stage.

In any case, this RAC rally entry had all been part of the master plan. Shortly afterwards the three Mk Is were shipped out to South America, and during the spring of 1970 they were used as recce and survey cars in that continent. 'Every time Culch came home from South America,' Browning quotes, 'he'd say that we needed to have a bigger aeroplane, more service, more wheels, more tyres or something.' Next Browning ordered six very special new monocoques from Pressed Steel at Swindon, which Triumph themselves paid for. Later Browning asked for more financial support, which was refused, and a clutch of rather frosty two-way correspondence about this has survived in the archives!

The first car to be completed – WRX902H – became a very hard-working test car, completing hundreds of gruelling miles at Bagshot and other test tracks, and later it would become a rally car for Brian Culcheth to drive on the Scottish rally.

Using this car as a guide, the World Cup specification eventually settled down. Enthusiasts trying to restore surviving cars in later years soon realized how special they actually were, for the body shells not only had foam-filled sills, many aluminium panels, flared wheel arches and engine bay vents in the front wings, but there was a great deal of extra strengthening, cool-air intakes in the roof, a bulge on the boot lid to allow a second spare wheel to be carried, and an aluminium roll bar.

The chassis spec, including a 150bhp engine, Stag-type gears in the gearbox, and a Salisbury Powr-Lok limited-slip differential, was conventional. Because of the huge distances to be completed the total fuel capacity of each car was no less than 32

Two pre-World Cup testing sessions saw all the works Triumph drivers involved. Paddy Hopkirk, Alice Watson and Rosemary Smith are in one group, Brian Culcheth, Brian Coyle, Andrew Cowan and Tony Nash in the other.

London to Mexico City, via Lisbon and Rio de Janiero, was no less than 16,000 miles (26,000km) for Daily Mirror World Cup rally *competitors. For the 2.5PIs, built like rallying tanks, it was an endurance test that was passed with flying colours.*

DAILY MIRROR WORLD CUP RALLY

gallons (145 litres), this being achieved by adding two Marston safety tanks to the normal 14-gallon (63.5-litre) tank.

One neat detail among many was the use of adjustable calibration for the fuel injection equipment, so that settings could be varied from sea level to the 16,000ft (5,000m) at which much of the Andes sections would be completed.

THE START

After months of preparation, it was almost a relief for the team when the World Cup rally actually started. Because there was no seeding on this event – starting numbers had actually been drawn at a glittering occasion held in London's Cafe Royal some months earlier – the works 2.5s were widely separated: no. 43 was Andrew Cowan/Brian Coyle/Uldarico Ossio; no. 88 Brian Culcheth/ Johnstone Syer; no. 92 Evan Green/Jack Murray/Hamish Cardno; and no. 98 Paddy

Hopkirk/Tony Nash/Neville Johnston. In addition, there were privately-entered 2.5PIs for no. 1 Bobby Buchanan-Michaelson/Roy Fidler/Jim Bullough and no. 39 A. Lloyd-Hirst/B. Englefield/K. Baker.

Buchanan-Michaelson's car was actually built up around a new works shell, and ex-works driver Roy Fidler had been persuaded to join the driving team. Both crews, however, were certain to keep begging for help, service or parts from the British Leyland team. Not only that, of course, but British Leyland also had to look after three Austin Maxis, whose lead drivers were H.R.H. Prince Michael of Kent, Rosemary Smith, and Flt Lt Peter Evans (of the RAF Red Arrows display team), along with a Mini Clubman for John Handley/Paul Easter.

Even before the event started Peter Browning admitted that his costs had spiralled above original estimates, so he was under enormous pressure to win.

Four works 2.5PIs started the World Cup rally from Wembley Stadium, but this car – driven by Evan Green, 'Gelignite' Jack Murray and Hamish Cardno – blew its engine in South America.

European Overture

The last few days were full of tension, but the crews were too busy to take too much notice. As Hamish Cardno, third man in 2.5PI no. 92 later wrote about the build-up:

> For two days the whole British Leyland team has been working at Abingdon – discussing the route and looking at route notes, writing pace notes, making detail modifications to the cars which have nearly driven the mechanics to distraction, posing for innumerable publicity photographs, choosing which spares we want, packing the car with them, and so on.
>
> Co-drivers (navigators in the case of three-man crews) are worried-looking people who sit huddled in the corner of the workshop or somebody's office surrounded by about three tons of paperwork. While somebody operates a power drill three inches from their left ear, they are trying to assess whether they'll have time to stop for a meal in Saarbrucken, or if they can stop for their driver to see his girlfriend at Munich....

It was only when they got to the start that the Triumph team, which had built what were effectively high-performance juggernauts, realized that their major rivals, Ford, had built strong but relatively lightweight Escort 'sports cars'. Most Triumph cars ran three-up, whereas Ford were planning two-man crews. Not only that, but four 2.5PIs were to battle with seven Escorts. Which philosophy would prevail after six weeks, particularly if the Triumphs had to overtake many of the ninety-six starters on the flat-out sections?

157

As far as the South American sections were concerned, Brian Culcheth, who so nearly won the event, once reminded me that: 'Basically, we always went as fast as we thought the car could stand. In Europe we had driven very gently, to keep in the frame. Then in South America two primes [special stages] were changed, which turned what should have been our advantage into a disadvantage – these were in Uruguay and Argentina.'

Lord Stokes, no less, turned up at the start to wish his team well. Peter Browning had no illusions about this: 'He expected us to win, not just to perform well. He was there to inspect his investment.' From the start, inside Wembley Stadium on Sunday morning, 19 April (where 'Gelignite' Jack Murray lived up to his name by letting off noisy firecrackers just before his car climbed the ramp. No one recorded whether Lord Stokes was amused or not), the cars had an easy three-day run through West and Central Europe, before turning back at Sofia, the capital of Bulgaria. From there to Monza, and the first overnight halt (after four days in a moving car!), the rally included the first two flat-out special stages (called 'primes' on this occasion). Evan Green's car was already in serious trouble with a misfiring engine, which was later traced to a faulty valve guide.

The first prime, a 50-mile (80km) blast through the Montenegran mountains of Yugoslavia (close to Titograd), was routine stuff, and saw Hopkirk's and Cowan's 2.5PIs just one minute away from cleaning the section. The second, a 100-mile (160km) section in the Serbian mountains, was sterner stuff; it did not help that some of the Triumphs actually had to stop for many minutes when they found a broken-down truck blocking the track!

After the night halt the cars moved across to the San Remo prime (cleaned by twenty-three cars), then tackled the famous French (Alpine rally) section from Les Quatre Chemins to Sigale. Although three of the 2.5PIs were fast here, the five-cylinder Green/Murray/Cardno car lost a front wheel 6 miles (10km) *before* the start, plunged off the road to rest against a tree, and

Early in the event, Brian Culcheth's Daily Mirror *2.5PI pauses at the Vienna control. He would go on to finish second overall, and be fastest on several of the flat-out primes.*

Dawn at Monza, during the Daily Mirror *World Cup rally. On the 2.5PIs, door plates have been covered over to meet Italian police requirements on the open road.*

stayed there for more than two hours until a tow truck plucked it out. Once extracted and bodged up, it tackled the prime with three brakes (one pipe had been nipped up), before being urged to the next BL service point at Rouaine where a massive rebuild followed.

By the time the injector for the faulty cylinder had been arranged to spurt petrol direct to the outside world (via a flexible pipe, which did not do much for the in-car ventilation on the left side!), by the time the suspension had been sorted out, and the chassis straightened, the car was hours behind the rally. Missing several passage controls, the stricken machine eventually reached Lisbon, but was well and truly out of the running. After the cylinder head had been changed, the car finally clocked in to board the *SS Derwent* a mere nineteen minutes before all access was refused. At that stage it was third from last, but determined to continue.

In the meantime, the three fit 2.5PIs drove across France and Spain to Portugal, before reaching Lisbon after seven days. Although Rene Trautmann's Citroen DS21 led the

event, with Hannu Mikkola's Ford Escort behind him, in spite of their delays in Yugoslavia the PIs were sixth, eighth and eleventh – and making up ground fast. Brian Culcheth's car was 27 minutes off the pace, and Hopkirk's 34 minutes, and both were blaming all of it on the hold-up in Yugoslavia.

There was still a long way to go, after all – as seasoned co-driver Henry Liddon pointed out when he showed that the first prime in South America would be longer than all the European Primes grouped together!

South America – the Longest Road Race Ever Held

All the World Cup cars spent nearly two weeks on the high seas before the *SS Derwent* arrived in Rio de Janeiro, so some of the crews had a rest. Most of them flew direct from Lisbon to Rio before spending a few days on the Brazilian beaches, but the most dedicated works drivers then rushed off to carry out last-minute practice on the Brazilian primes.

After seventy-one cars restarted from Rio

on 9 May, they faced a twelve-day slog around the South American continent (and about 9,000 rough and tough miles (14,500km)), with night stops located only in Montevideo (Uruguay), Santiago (Chile), La Paz (Bolivia) and Lima (Peru), before the surviving machines boarded another boat which would take them from Buenaventura in Colombia to Panama in Central America.

This was where Peter Browning's philosophy was going to be proved or blown out of the water. Because the 2.5PIs had been built like tanks, and mostly carried three-man crews to help keep fatigue to a minimum, they probably could not be fastest on the primes unless these were very rough. On the other hand, they could possibly turn out to be the most reliable. This marathon, they hoped, would be won on stamina, determination and organization, with 'Browning's bomber' – the British Caledonian Bristol Britannia aircraft, fully loaded with spares and hopping from halt to halt – as part of that approach. Ford's 'You bend 'em, we'll mend 'em' approach, with spares dumps in each country, was utterly different.

British Leyland's strategy was based on road and track conditions found on the extensive reconnaisance trips, where some of the unmade sections were often very rough. To their horror, when they reached the Uruguayan and Argentinian primes the Triumph crews discovered that proud local authorities had carefully graded every stretch to be tackled by the event:

In Argentina the time schedule was easy, [Culcheth recalls] compared with a difficult recce, because the authorities had regraded all the routes – they were like a motorway. In Uruguay it was similar. Andrew Cowan and I were first and second fastest – Andrew by a minute, but I'd had to stop to change two punctures – and we beat the Fords by several minutes. I seem to recall that we *averaged* 108mph (174km/h) on that prime.

In Argentina I reckon we could have been an hour faster than the Fords if we'd all been penalized, because they had a very limited top speed, whereas we could do 125mph (200km/h).

Although Evan Green's car reached Argentina, it finally retired when the long-suffering engine threw a con-rod and ventilated the cylinder block: not even British Leyland's mechanics could sort that one out.

When the forty-two surviving cars left Santiago in Chile, finally turning north towards Bolivia, Peru, Ecuador and Colombia to catch the boat to Panama, the 2.5PIs were in fourth (Culcheth), seventh (Hopkirk) and tenth (Cowan). Unhappily, on the incredibly long (510 mile/820km) rodeo – La Vina 'Gran Premio' prime in Argentina, which the fastest cars completed in 9–10 hours, Andrew Cowan's car crashed, with near fatal consequences to the crew.

By chance, I was running the time control at the end of this prime, and when news finally filtered through to us, everything sounded grim. Cowan, making remarkable time, had caught and passed many cars, but only 27 miles (43km) from the finish he had been blinded in a dust cloud when catching another car, and gone off the road. The car was written off, its roof totally flattened. Cowan and co-driver Brian Coyle were knocked about, but Lacco Ossio suffered crushed vertebrae in his spine and a broken skull, all of which need immediate surgery. It took time, but happily all of them made a complete recovery.

More ultra-long primes, some of them running at an altitude of more than 15,000ft (4,500m), saw the field reduced to a mere twenty-six cars before they were loaded on the Italian line ship *Verdi*. By this time Culcheth had clawed his way up to third place, an hour behind the leading Ford. Although Hopkirk's 2.5PI was fourth, this was in spite of a big shunt in Ecuador, after

This relaxed shot of Brian Culcheth's World Cup *2.5PI, which finished second overall, shows off all the Abingdon detail applied to the building of these cars, including the cool air vent in the roof, the engine bay vents in the wings (behind the front wheel-arches), the extra ventilation scoop ahead of the base of the screen, and the combination of sump guard mountings/fog lamp supports.*

a rear brake pipe had split. In my report on the rally in 1970, I noted that: 'The big Triumphs are very strong, and after a front suspension change, a new radiator and a bit of bodging, the car felt almost like new.'

On any other rally, a final 51 hour 30 minute section to the finish would have seemed gruelling, but on the World Cup this Panama City–Fortin leg looked like a doddle. It was eventful for the Triumphs, for Paddy Hopkirk's car set fastest time on one of the primes, the second-place Ford went off the road, and suffered a flattened fuel-supply pipe; all this cost more than 90 minutes, and the result was the Brian Culcheth's 2.5PI, still unmarked after its six-week ordeal, moved up into a magnificent second place.

Outside the Olympic stadium in Mexico City, the team celebrated second and fourth places overall. To quote Peter Browning: 'Nothing went wrong. The service plan worked, the cars worked. Paddy's car was handicapped by overweight by being three-up. If Paddy had been two-up, who knows? He might have outpaced Brian.' Afterwards *Motor*'s rallying expert, Hamish Cardno, commented that:

British Leyland were too cautious. Their most likely potential winner, the Triumph 2.5PI, was less nimble than the Escorts because of its size, but on those primes where sheer straight-line performance was an advantage its higher top speed paid off. However, excessive caution on the part of two of their best three crews cost them dearly . . . Brian Culcheth stuck his neck out and reckoned it was worth trying two-up – a decision which must make his second place even more satisfying.

Basically, the Triumph has emerged from the event as a very good car. Apart from fairly frequent replacement of front struts and shock absorbers – components put under great stress by the weight of the cars and their crews – they gave no trouble . . .

Even so, not everyone was delighted by this performance. Quite inexplicably, Lord Stokes thought the team effort had been a failure – he simple could not understand the enormity of what had been achieved. Peter Browning's memory is that:

It was our failure to win the World Cup, I think, which signed the death warrant of the department. Second in the London–Sydney marathon of 1968 (in a BMC 1800) had been pretty good, *but we hadn't won.* I think that losing again was too much for him to bear, especially after all the bills came in, and we had only finished second.

Certainly when the surviving cars came back to the UK they were quickly sidelined. Neither turned a wheel in anger for several months, and they spent most of their time in store at Abingdon, gathering dust.

THE END OF TRIUMPH RACING

As soon as he got back from Mexico City, Brian Culcheth prepared to tackle the Scottish rally, the event which had been so frustrating for the 2.5PI a year earlier. The old World Cup test car was not only refurbished, but lightened by more than 400lb (180kg) and was given a more powerful engine. According to Culcheth:

It was a lot more powerful than a standard rally car with about 20 extra bhp. At that time Triumph North America were getting a tremendous power output from their PI engines, supposedly over 200bhp, but the Weber engine we used had about 165bhp at the flywheel [I think this is an

Immediately after the World Cup *rally, Brian Culcheth and Johnstone Syer flew home, picked up the refurbished test car, and won the Scottish rally outright. WRX902H was much lighter and considerably more powerful than the 'tanks' Abingdon had built for the* World Cup *rally.*

exaggeration], whereas the World Cup cars gave about 138bhp at the same point. The Scottish car also had those marvellous gear ratios.

The Scottish car, in fact, was a lot quicker than the World Cup cars, which were a lot heavier – in them it was only the downhill bits which were exciting.

As expected, the Scottish was hot, dry, rocky and fast. Competition came from Roger Clark's works Escort RS1600, Paddy Hopkirk's special Mini Clubman 1275S and Harry Kallstrom's Lancia Fulvia HF, so in the early hours the big Triumph was a few places off the pace.

After 36 hours Culcheth was third behind Clark and Kallstrom, but on the fourth day the Ford's new-fangled BDA engine gave up the ghost. On the last day it was also revealed that Kallstrom's Lancia had been running very late on the road (having stopped for spares and service to be provided) and if regulations, and their interpretation, were to be believed, this penalized him heavily.

Culcheth's Triumph, therefore, moved smoothly into the lead, comfortably ahead of Hopkirk's Mini, but it would be weeks before the Lancia's protests were finally thrown out and the 2.5PI's splendid victory confirmed.

Immediately afterwards, though, the rundown of the British Leyland (for which read Triumph) rallying effort began. Peter Browning, asked to make recommendations for a future strategy, soon realized that British Leyland would not back the development of a new 'homologation special'. Further, 'I don't think we could have made the 2.5PI any better. I think it was a super "rallying tank" for marathons and safaris, but I don't think we could have made it an outright winner in other events.'

Brian Culcheth thought otherwise :

I thought there was a tremendous amount that could be done. The 2.5PI had all the

potential of the Peugeot 504s, which were similar products in many ways. It could have been so right for all the rallies the 504s went on to win. I was really looking forward to 1971, for Peter was hoping to do the 1971 safari with World Cup-spec cars . . .

In the end, British Leyland closed down the famous Abingdon competitions department in October 1970. Officially this meant an end to the 2.5PI's rally programme, but Brian Culcheth, the dedicated professional, wanted to stay in the sport:

I was desperate to keep going, and said so to the public relations staff at British Leyland, who included Simon Pearson and Mike Greasley. They must have mentioned this to Bill Davis, who was Triumph's top director. He called me to a meeting, where I found that he was particularly keen to continue Triumph's association with sport. But there wasn't any money.

Brian Culcheth-Team Castrol

After Culcheth had cast round for sponsorship, getting help from Castrol, the oil company, and being retained by British Leyland International for promotional and demonstration work, the result was the birth of the Brian Culcheth-Team Castrol operation. Apart from the occasional loan of service vehicles from Abingdon (where Special Tuning was now in the driving seat), its sole piece of machinery was Culcheth's ex-World Cup 2.5PI – XJB305H !

Even so, this project sounds grander than it actually was. Culcheth got an office in British Leyland House, which was then in Marylebone Road, London, shared secretarial services, spent hours on the phone, mined small tranches of money from local BL sources, and did a handful of events in the 1971 season.

It was by no means a busy year for him, for his first event was the Welsh International,

Testing, testing . . . For 1971 Brian Culcheth took over his ex-World Cup 2.5PI, set up Brian Culcheth-Team Castrol, and tackled events as a semi-private owner. With Triumph tester Gordon Birtwistle alongside him, he is here testing the rejuvenated car early in 1971.

in May. By this time the famous ex-World Cup 2.5PI was somewhat lighter and simpler than before – the scrutineers even saw fit to reject it at first because of an inadequate roll cage! The Welsh was not a happy event for the team, as the car suffered fuel feed problems, and could only record fourteenth place.

Then came the Scottish, an event which Culcheth wanted to win again. For 1971 his PI was neither as light nor as powerful as the 1970 example, but it was still competitive on the long and rocky stages. Fourth at one stage, it eventually finished tenth after two off-road excursions.

Then, in September, came the only overseas event with this car, the hot and rough Cyprus rally, where the 2.5PI was ideally

suited to the surfaces, but not for the twisty nature of the stages. All the same, second place overall was a fine result.

That, however, was the end of the road for this gallant car (it is one of the few ex-works big Triumph saloons that survives to this day), though it was not quite the end for the 2.5PI. In previous years privately prepared 2000s had performed well in the East African Safari, though they had never been fast enough to win: for 1972 the Triumph importer to Kenya – Benbros – found enough finance for one last works entry, a 2.5PI for the 1972 Safari.

This was a brand-new car, assembled at Abingdon by Special Tuning, and including many recognizable 'World Cup' features. Culcheth himself was very enthusiastic, but

Brian Culcheth is still convinced that a properly tested and financed team of 2.5PIs could have won the East African Safari. Driving a singleton entry in 1972, he managed to finish and win his class, but the structure was in a sorry state at the end of the event.

because the car only just arrived in Kenya in time to take the start, he was neither relaxed, nor properly prepared. He reminds me how things went from bad to worse on the event itself:

One thing I had discussed with Bill Davis at this stage was the potential of the 2.5PI in the Safari but we had a rally of considerable mechanical problems. It started only 100 miles [160km] after the start when an antelope went into the front of the car, and the horns went straight through the radiator. There was water gushing out everywhere. I was with Lofty Drews, a local co-driver, so we got some mud and tried to cake it up, but that didn't work properly. Then we saw a privately owned Triumph 2000 sitting in a drive, and you can guess the rest! We begged the radiator from the owner, and changed it.

Later in the event the gearbox stripped most of its gears (nothing new there, if you recall the mid-1960s), and after the half-way halt

the rear screen fell out, just as it had done several times in pre-World Cup testing, and finally the whole of the rear end of the body shell started to break away from the cabin. By the finish the boot looked as if it was ready to fall off – Culcheth and Drews were lucky to finish at all, in thirteenth place.

After that disappointment (surely British Leyland could not have expected the car to win, especially when the event was being dominated by 220bhp Escort RS1600s?) the 2.5PI programme was wound down completely, and for the next few years Culcheth divided his time between rallying Morris Marinas, and developing the smaller Triumph Dolomite Sprint model.

Looking back after more than twenty years, it is easy to suggest that the 2.5PIs might have done better if British Leyland had believed in them, and backed them with more facilities and larger budgets. Maybe that is so, but I doubt if they could ever have been consistent winners, for this was the period when the Ford Escorts – Twin-Cams and RS1600s – were in the ascendant, when

165

rear-engined cars like the Alpine-Renaults and Porsche 911s were so effective, and when the free-spending Fiat-Lancia organization was producing a whole series of special cars.

The 2.5PI as a rally winner, then? It is a nice thought, but the cars probably achieved all that could be expected of them.

Year/Month	Event	Crew	Result
International rallies – the 'Injection' years			
1967			
November	RAC	D. Hulme/G. Robson R. Fidler/A. Taylor	Event cancelled only hours before the start
1969			
May	Austrian Alpine	P. Hopkirk/A. Nash	DNF
June	Scottish	B. Culcheth/J. Syer	2nd in class
November	RAC	A. Cowan/B. Coyle	11th overall, 1st in class
		P. Hopkirk/A. Nash	2nd in class
		B. Culcheth/J. Syer	3rd in class
1970			
April/May	World Cup	B. Culcheth/J. Syer	2nd overall
		P. Hopkirk/A. Nash/ N. Johnston	4th overall
		A. Cowan/B. Coyle/ U. Ossio	DNF
		E. Green/J. Murray/ H. Cardno	DNF
June	Scottish	B. Culcheth/J. Syer	1st overall
1971			
May	Welsh	B. Culcheth/J. Syer	14th overall
June	Scottish	B. Culcheth/J. Syer	10th overall, 2nd in class
Sept.	Cyprus	B. Culcheth/J. Syer	2nd overall
1972			
April	East African Safari	B. Culcheth/L. Drews	13th overall, 1st in class

9 The 2500TC and 2500S Models

Soon after it was formed in 1968, British Leyland stated that it had thoroughly reviewed every aspect of its business, and that a long term strategy was already in place. Fine words, but as far as most of us can now see, it simply was not true. Until 1972, at least, if there *was* a new-model strategy, no one explained it to the constituent companies, which were left to fight among themselves.

Because Standard-Triumph had been the first car-builder to be drawn into Leyland's net, it was always assumed that Triumph would be the favoured marque in British Leyland. It certainly looked like that at first. For years Triumph was left to carry on its battle against Rover – though the two companies were both owned by British Leyland – and there was strictly no co-operation between the two concerns.

In the early 1970s, as in the 1960s, the Triumph 2000 was left to slug it out against the Rover 2000. This was *laissez-faire* economics, and purely in business terms made no sense at all. Whereas British Leyland should have been looking for big economies by trimming competitive ranges, they actually allowed things to go the other way. Four years after British Leyland had been set up, not only were both the 2000s still on sale, but both had been face-lifted, and more variations of each type were available.

My belief is that if British Leyland had not been formed, and if the art of 'Product Planning' so assiduously developed in this country by Ford, had not spread throughout the motor industry, I doubt if the final derivatives of Innsbruck – the Triumph 2500TC and 2500S models – would ever have been put on sale. They only appeared because Triumph and Rover were finally obliged to merge their interests into the Rover-Triumph combine, and because British Leyland opted for one major new model – the SD1 (Rover 3500) – to be designed and developed.

Because SD1 was a new design that was to be built in a new factory on which work had not started, it was going to take a long time to get the new model ready for sale. In the meantime, every advantage of existing Triumph (and Rover) models had to be taken.

Under British Leyland, indeed, the Triumph 2000 family had already led a charmed life (and a profitable one) for several years. Because the Innsbruck restyle had already been approved before the formation of British Leyland upset many priorities, it surged safely into production in the autumn of 1969, and even in the early 1970s the showroom battle between the two 2000s – Triumph and Rover – was allowed to carry on unabated.

THE ADVENT OF PRODUCT PLANNING

As the 1970s opened Bill Davis, Triumph's managing director, invited an ex-BMC colleague, Alan Edis, to open up a product planning office at Canley, for like every other manager in the corporation he was under pressure to increase sales and market share.

By 1972, after Rover-Triumph had been formed and Mike Carver had taken over a combined product planning office, the British car market was booming, and Lord Stokes thought both companies could make more of it. Although Edis knew that the Rover and Triumph 2000s, now theoretically co-operating rather than competing, would have to soldier on for years yet, he was asked to look at ways of freshening up established ranges.

Immediately after the Rover-Triumph business was established, the table below shows how the *combined* ranges stacked up against each other.

To Edis and his staff, this list showed up a deficiency and an opportunity. There was a large gap – £218, or 13.4 per cent – between the 2000 and the 2.5PI, which needed to be filled. On the 'wouldn't it be nice if . . .' basis, a Triumph gap-filler with about 100/110bhp would be ideal. But could it be done? And quickly?

'I thought that was an obvious hole in the range,' Edis recalls, 'and Bill Davis backed me. I think the customers were looking for a car like that, and I think the drivability of the 2000 was much improved by having a larger capacity engine. It was an easy job to do.'

A small number of 2.5-litre, carburettor-engined cars had already been produced for supply to South Africa, so much of the engineering had already been done. Suddenly, this problem also turned itself into an opportunity. By the early 1970s Triumph was becoming thoroughly irritated with the quality and reliability failings of the Lucas fuel-injection system, and was inclined to turn its back on the installation. (In any case, none of their major rivals had matched the injection system, so Triumph thought that there would be no loss of face if it was abandoned . . .)

The opportunity, therefore, was not only to plug the existing gap between 2000 and 2.5PI but, if it was carefully done, to provide an acceptable alternative to the 2.5PI as well.

Although Edis's staff put in a great deal of time to this programme, Edis himself dismisses it briefly and succinctly:

> We introduced the carburettor version of the 2500 in order to spread price alternatives, and to give nicer driving characteristics. We did look at some front-end face-lift action, but that wasn't introduced in the end. Most of our concentration, of course, was on the new car – and, to be fair, about equal emphasis was given to the Triumph *and* Rover 2000s in this period. There were other priorities – the Dolomite and Dolomite Sprint for instance.

Incidentally, there was also a short-lived proposal to put the Stag-type V8 engine into the Triumph 2000 structure, but although a car

Model	Engine (litres)/power (bhp)	UK Retail Price
Triumph 2000	**2.0/84 (DIN)**	**£1,626**
Rover 2000SC	2.0/91	£1,723
Rover 2000TC	2.0/114	£1,829
Triumph 2.5PI	**2.5/132 (DIN)**	**£1,844**
Rover 3500S (manual)	3.5/147	£2,026
Rover 3500 (Auto)	3.5/147	£2,104

At the same time as the 2500TC was launched in 1974, Triumph also retouched the 2000. The cars shared these wheels until the end in 1977.

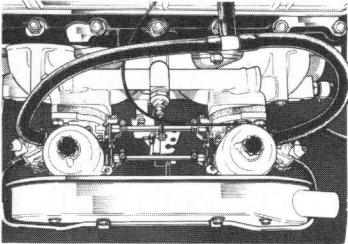

The 2500TC was introduced in 1974, as a half-way house model between the 2000 and the 2.5PI. In effect this was a 2.5PI engine with the 2000's camshaft profile and carburation. By this time SU carburettors had replaced Zenith-Stromberg ones on all Innsbruck models.

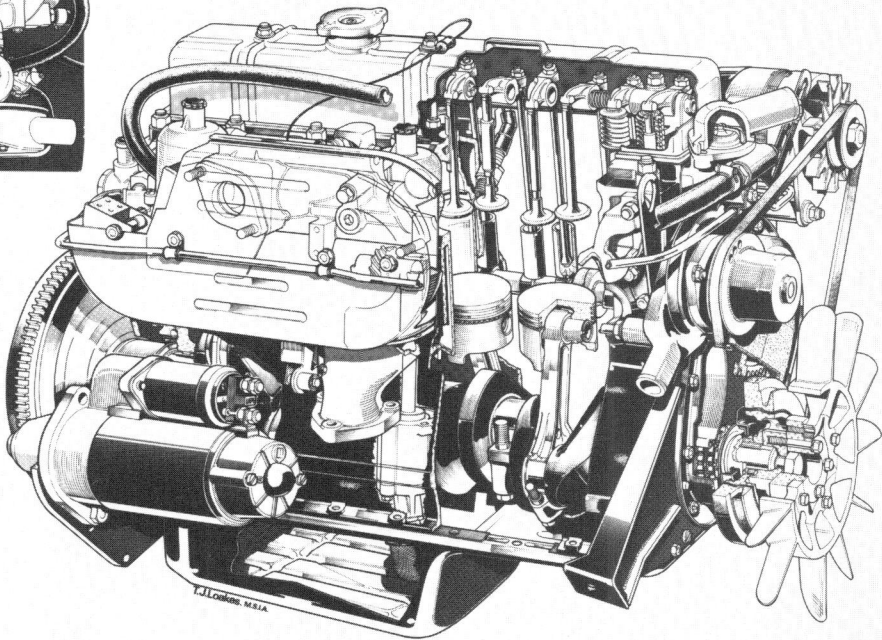

169

was built for sales director Lyndon Mills to use, nothing further came of this.

The strategy, therefore, was to develop not one but two different versions of the larger-engined six-cylinder engine, and eventually to allow the PI derivative to die a natural death. The first 2.5-litre engine to be put on sale would be a quick and simple 'parts bin' job, the second would involve more fundamental development. However, 'I don't feel we abandoned Lucas fuel injection,' Edis told me. 'That's the wrong word. More sophisticated and better injection systems came along, and in a sense when we introduced the 2500S, that was better value for money. We offered the carburettor version that offered people very nice driving characteristics.'

And so it was. Starting in 1972, with the first production cars being built in December 1973 and February 1974, and public launch following in May 1974, this development programme seemed to take very little time. In typical Triumph copybook manner, production of the first new car, called 2500TC, began in earnest in March 1974. The 334 cars in that month were followed by 967 in April, after which assembly became a matter of routine.

2500TC – THE NEW PACKAGE

Except that there was a new slatted front grille (of Stag type) and badges to tell their story, along with rubber inserts in the bumper blades, and a Stag-type steering wheel, Triumph made virtually no style changes for the new version of the car. As expected, saloon and estate car types were made available. Indeed, the intention was always to market this car as an up-engined 2000, no more and no less.

The engine itself used the same bore, stroke and modified cylinder block of the 2.5PI, while the cylinder head of the engine, was effectively a lower-compression ratio

version of the 2.5PI unit, fitted with the same type of manifolding and twin carburettors (SU, not Zenith-Stromberg, by this stage) as the 2000.

It was, in fact, a splendid example of Triumph's long-established mix-and-match abilities. The USA-market TR6 sports car also used a 2,498cc engine with twin carburettors, and except for the use of SUs instead of Zenith-Strombergs in the 2500TC, the two units had a great deal in common.

Thus modified, the 2.5-litre six-cylinder engine produced 99bhp at 4,700rpm, with a peak torque of 133lb/ft at 3,000rpm. This, together with the use of a much higher final drive ratio (3.45:1 instead of the 2000's 4.0:1) seemed to make a significant difference to the drivability of the new model.

The Right Car at the Right Time

With very little effort, therefore, Triumph had converted the Innsbruck range to six separate models, just as the need for more economical cars became apparent. When deliveries began in mid 1974, the table below shows how specifications and power outputs stacked up. By this time, incidentally, optional power-steering cost £90, and overdrive (which most cars seem to have) £99.50.

Although the 2500TC was launched at the worst possible time in Europe, when the continent was recovering only nervously from

Model	Engine (litres)/power (DIN bhp)	UK Retail price
2000 Saloon	2.0/84	£2,051
2500TC Saloon	2.0/99	£2,166
2500PI Saloon	2.5/120	£2,380

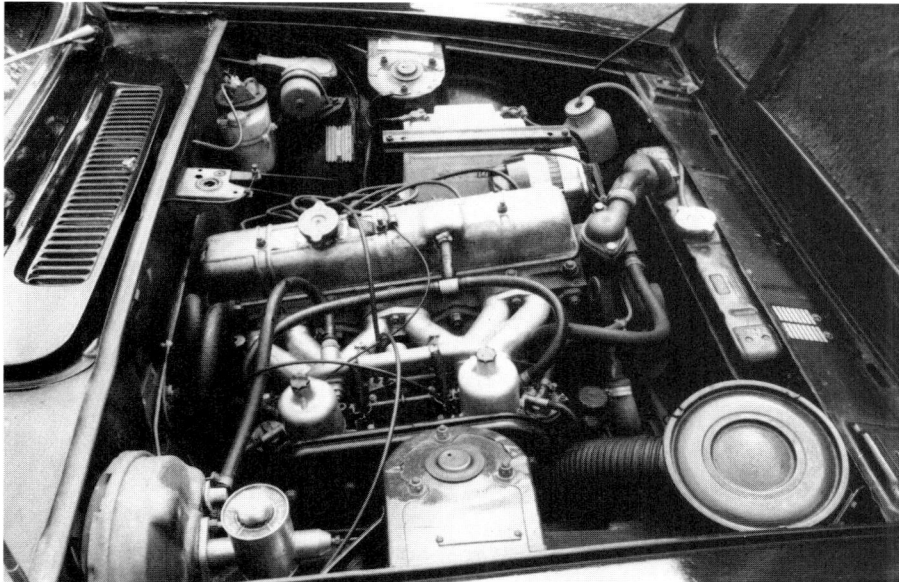

Visual differences between 2500TC and 2500S engines were small but significant. The original 2500TC of 1974 had 99bhp, and its inlet manifold had a balance pipe near the cylinder head face. The 106bhp 2500S engine of 1975 had a new, more curvaceous inlet manifold, with the balance pipe closer to the enlarged SU carburettors. To maintain clearance, the carburettor air box of the 2500S was very slim indeed.

Triumph 2500TC and 2500S (1974–7/1975–7)

Produced
CKD kits first sent to overseas assembly plants in 1972. 2500TC: December 1973 to May 1977 (Uprated engine spec from March 1975). 2500S: March 1975 to May 1977

Identification
Chassis numbers carried the prefix MM (2500TC) and MP (2500S). CKD 2500TC cars carried the prefix MK. All mid-1975 modifications were phased in with introduction of the 2500S model

Layout
Unit-construction body/chassis structure in steel. Five-seater, front engine/rear drive, sold as four-door saloon, or five-door estate car.

Engine
Type	Standard-Triumph six-cylinder
Block material	Cast iron
Head material	Cast iron
Cylinders	6 in line
Cooling	Water
Bore and stroke	74.7 x 95.0mm
Capacity	2,498cc
Main bearings	4
Valves	2 per cylinder, pushrod and rocker operation
Compression ratio	8.8:1 (from 1975 8.5:1)
Carburettors	2 SU HS4 (from 1975, 2 SU HS6)
Max. power (DIN)	99bhp @ 4,750rpm (from 1975 106bhp @ 4,700rpm)
Max. torque	133lb/ft @ 3,000rpm (from 1975 139lb/ft @ 2,750rpm)

Transmission (Manual)
Clutch	Single dry plate, 8.5in diameter; diaphragm spring, hydraulically operated

Internal gearbox ratios
Top 1.00, 3rd 1.386, 2nd 2.100, 1st 3.28, reverse 3.369
Final drive 3.45:1
19.9mph/1,000rpm in direct top gear
Optional Laycock overdrive (on top and third gears) had a ratio of 0.797:1, overall ratio 2.75:1. Overdrive standardised from mid-1975.
25.0mph/1,000rpm in overdrive top gear

Automatic transmission (optional)
Torque converter Maximum torque multiplication 2.0:1

Internal transmission ratios
Top 1.00, intermediate 1.45, low 2.39, reverse 2.09
Final drive 3.45:1.
19.9mph/1,000rpm in direct top range

Suspension and steering

Front	Independent by coil springs, MacPherson struts, lower wishbones, telescopic dampers in struts. Anti-roll bar from mid-1975
Rear	Independent by coil springs, semi-trailing wishbones, telescopic dampers
Steering	Rack and pinion (optional power assistance, standard on 2500S model)
Tyres	(2500TC) 175x13in radial-ply; 185x13in radial-ply from mid-1975 (2500S) 175 × 14in radial-ply
Wheels	(2500TC) Pressed steel disc, four-stud fixing (2500S) Cast alloy disc, four-stud fixing
Rim width	(2500TC) 5in (2500S) 5.5in

Brakes

Type	Disc brakes at front, drum brakes at rear, with vacuum servo assistance
Size	9.75in diameter front discs; 9 x 1.75in wide rear drums

Dimensions (in/mm)

Track	
Front	52.5/1,333
Rear	52.9/1,344
Wheelbase	106/2,692
Overall length	182.3/4,630 (saloon) 177.25/4,502 (estate)
Overall width	65/1,651
Overall height	56/1,422
Unladen weight	(saloon) (2500TC) 2,681b/1,216kg (2500S) 2,696lb/1,223kg (estate) (2500TC) 2,801lb/1,270kg (2500S) 2,842lb/1,289kg

the effects of the first energy crisis, and *all* car sales were badly hit, this was a strategy that seemed to work well. It might have been pure luck that a more fuel-efficient big Triumph arrived at the same moment that fuel prices had started to soar, but Triumph never admitted to that. At launch, the company claimed that the new version was designed 'specifically to meet the current need for greater fuel economy'.

In product planning terms, the 2500TC caused quite a lot of 'substitutional buying' – sales were being won from customers who might otherwise have bought 2000s or 2.5PIs – but there was no doubt that it was what the markets wanted. By the end of 1974 the 2500TC had already become the best-selling version of the range. No fewer than 9,618 2500TCs were made in an annual total of 21,828 cars – 44 per cent of the total.

Closer examination of the figures showed that the 2000 itself had been affected by the arrival of the 2500TC, but that the 2.5PI had taken a very big hit indeed: 12,934 PIs had been sold in 1973, but only 3,452 followed them in 1974. It was, indeed, the beginning of the end for the fuel-injected cars, which would disappear altogether in 1975.

Because the 2500TC was better, but not sensationally better, than the 2000, road test reports showed approval but not astonishment. *Autocar* thought the 2500TC to be rather expensive, though admitting that it had 'better economy. Reasonable performance. Ride and optional power steering good; roadholding adequate. Pleasing finish, comfort and equipment . . .'

Even so, this was a car that reached 100mph (160km/h), with 0–60mph (0–100km/h) in 11.8 seconds, and was therefore significantly quicker than the 2000 from which it was developed. Most significant, though, was the comment that: 'Unlike some test 2.5PI Triumphs, this carburettor car idled smoothly without the lumpiness of the fuel-injected engine.'

Motor was equally friendly, but unruffled, summing up its verdict of the car as: 'On the whole sound, comfortable and agreeable transport, though it may not inspire the discerning enthusiast.'

But hindsight is a wonderful thing. Looking back, it is easy to say that the 2500TC should have been developed years earlier, and that Triumph could have sold more cars. Even if not for the last part of the Mk I's career, it could certainly have been ready for the launch of the Mk II. By that time the 2.5-litre engine had been in production for two years, for fuel-injected *and* carburetted versions had both been fitted to the TR5 sports car from the autumn of 1967.

On the other hand, 2.5PI sales might have melted away much quicker than they did. In any case the records show that 2.5PI Estate car production was run down very rapidly in the spring ahead of the arrival of the 2500TC Estate.

THE 2500S

The scene was now set for the Mk II range to receive its final reshuffle. Development of the new Rover SD1 model was proceeding well, and Rover-Triumph still had every hope of announcing all versions of the car during 1976, which meant that time was running out for both the Triumph 2000 range and the Rover P6 range.

However, with two selling seasons still to go, the planners clearly thought that there was room for one more series of changes to the ageing Triumph range. By the winter of 1974–5, in any case, the market place had begun to make its own decisions.

Sales of the fuel-injected cars seemed to have gone into terminal decline, for 2.5PI production was down to a trickle (none were built, in fact, during February, March and April 1975), but Triumph wanted to replace that car with a more powerful carburetted version of the engine. Well before that car, called the 2500S, was introduced in June 1975, the 2.5PI had been condemned. Only 416 saloons – and no estate cars – were produced in 1975, the last one of all coming off the line early in July.

Even though British Leyland plunged into near bankruptcy in January 1975, when the British government moved in with a rescue and subsequent nationalization, plans laid for the 2000/2500 range in 1973–4 went ahead. For the final two years, there would still be three slightly different types of 'big Triumph', but this time all of them would share the same type of SU-carburetted engines. A new model, the 2500S, would be the most completely equipped 2000-type car of all time.

A NEW LINE-UP

Although the new line-up made a great deal of marketing (and production) sense, it no longer seemed to be as exciting as before. But then, in cash-strapped British Leyland, what was?

Instead of 2000, read 2000TC and instead of 2.5PI, read 2500S. In addition, there was a 2500S estate car derivative, but all other

The way to identify the 2500S of 1975–7 was to note its 14in diameter spoke alloy road wheels . . .

. . . with the exposed four-stud fixing. Originally these cars had 175-section tyres, but a later owner has fitted 185s on this particular car.

A selection of new badges on the nose, tail and rear quarter panels all helped to identify the 2500S of 1975–7, which was extremely successful, in fact a much-needed minor triumph (no pun intended) for British Leyland's product planners.

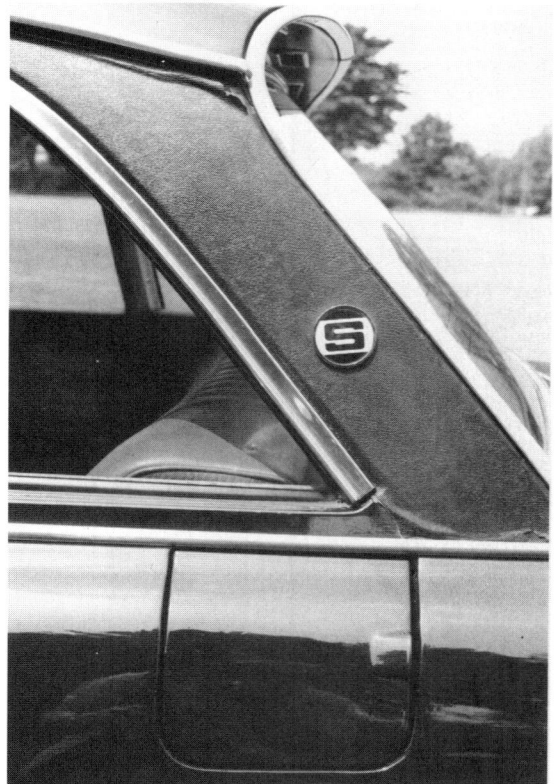

estate car types were dropped. The 2500TC and 2500S models shared the same engine and transmission, but there were other differences to signal the higher specification of the 2500S. This sounds simple enough to understand – except that a great deal of detail change was hidden away under the unchanged styles (*see* table below).

Model	Engine (litres)/power (DINbhp)	UK Retail Price
2000TC Saloon	2.0/91	£2,713
2500TC Saloon	2.5/106	£3,076
2500S Saloon	2.5/106	£3,271
2500S Estate	2.5/106	£3,742

In summary, power was up for all models, but so were prices, though the increase in costs could mainly be blamed on the rampant inflation that afflicted the country at this time. The extra cost of overdrive, standard on both the 2.5-litre models, did not help.

Although there was virtually no change to the cars' exterior styling, and very few inside the cabin, the engineers had made many improvements to the engine, transmission and suspension.

Engine

By slightly relocating the engine installation – the entire assembly had been moved over 0.6in (15mm) to the left side (the 'battery side') of the engine bay – more space had been found for the longer and more efficient inlet manifolding of the USA-market TR6 engine, and for larger (SU HS6 type with 1¾in chokes, instead of HS4) carburettors. This had been eased by altering the shape of the air inlet box to the carburettors. At the same time, a larger-diameter exhaust system, the use of a viscous coupling to the cooling fan, and revised valve/camshaft

Although both SU (shown here on a 2500TC) and Zenith-Stromberg carburettors operate on the constant vacuum, sliding needle principle, they are very different indeed in construction. In the big Triumphs, SUs had taller dashpots and a large piston hidden inside their aluminium castings.

177

Model	Peak power and peak torque
2000TC (2000)	84bhp at 5,000rpm (91bhp (DIN) at 4,750rpm) 100lb/ft at 2,900rpm (110lb/ft at 3,300rpm)
2500TC and 2500S	106bhp at 4,700rpm (99bhp (DIN) at 4,750rpm)
(1974–5 2500TC)	139lb/ft at 2,750rpm (133lb/ft at 3,000rpm)

timing all helped to provide more power and torque (*see* table above).

Not only did the new engines produce better power and torque, but they were also more flexible than ever before (Stanley Markland's dictat of 1961–2, about top gear acceleration from 10mph/15km/h, had never been forgotten).

Transmission

To take advantage of the improved engines, Triumph also raised the overall gearing on the 2000TC, for the new rear axle ratio was 3.7:1 (it had been 4.1:1 on the obsolete 2000). This meant that for the 2000TC the gearing equalled 18.6mph per 1,000rpm (29.9km/h per 1,000rpm) in direct top gear, instead of 17mph per 1,000rpm (27.3km/h per 1,000rpm) for the 2000.

As before, overdrive was an optional extra on the 2000TC, but standard on the 2500TC and 2500S. Borg Warner automatic transmission, still an expensive extra, was only bought by a small percentage of customers.

Suspension

Looking back, it seems amazing that

Little boys were always impressed by the 140mph (225km/h) reading speedometer on 2.5PIs and 2500S models, even though neither car could reach 110mph (177km/h)when flat out.

Once the 2500S was introduced, the range was rationalized, with the estate car body shell only available in 2500S tune. Like all previous estate types, it had a neat, lockable cap for the fuel filler on the right flank.

Triumph did not include a front anti-roll bar in the specification of the 2000 range until 1975. Even though they always strove to provide a soft limousine-like ride, for the first twelve years of the cars' life this was achieved by stiff front springs and firm front strut settings.

A lot of work having gone into producing sporty handling for the related Stag model, this finally benefited the saloons as well. At last, for the 2000TC/2500TC/2500S line-up, a 1.25in (32mm) anti-roll bar was specified, along with different strut settings. Radial-ply tyres had been standardized for some time, but the 2500TC was now given fatter (185-section) radials, while the 2500S not only had 175SR-14in tyres, but these came on Stag-type alloy wheels with a 5.5in rim width.

All this, complete with power-assisted steering as standard equipment, and a smaller-diameter steering wheel, gave the 2500S a much more level ride than ever. Edward Eves, when describing the new car in *Autocar*, stated that it 'now rides and steers like a good sports car; moreover it corners like one'. Perhaps that was going a little over the top, but it certainly gave a flavour of the car's new-found agility.

Equipment

Very few changes were made to the 2000TC and 2500TC models, but the 2500S was an appealing package. Apart from its five-spoke Stag-type alloy wheels and the '2500S' badges, there was a vinyl-trimmed rear quarter panel, and a matt-black colour across the vertical panel on the tail. Inside the cabin, the leather-covered steering wheel was accompanied by an instrument panel layout based on the obsolete 2500PI, this time with separate rev counter, while front-seat head-rests were fitted as standard.

Performance

When *Autocar* tested a 2500S in July 1975 (the test car, JHP368N, was the same used by Rover-Triumph in the new model announcement several weeks earlier), it recorded a top speed of 105mph (169km/h) in overdrive *and* direct top gear, along with 102mph (164km/h) in overdrive third, which asks several pertinent questions about the car's gearing and its aerodynamic qualities.

The 0–60mph (0–100km/h) acceleration sprint in 10.4 seconds made this the quickest car of this type ever tested – faster, even, than the 2.5PI that the 2500S replaced, though the 2.5PI was faster up to 100mph (160km/h). A look at all the individual acceleration times in a particular gear showed that the 2500S was as punchy as the 2.5PI in all respects. Yet the 2500S had 106bhp, and the 2.5PI was claimed to have 120bhp: the gearing, and the gear ratios, were effectively the same, so what conclusions are to be drawn from that?

The most penetrating remarks in this test were reserved for comments on the handling.

The improvement wrought by the massive front anti-roll bar is so marked that one wonders why it was not done before, and why it has been left off the other two cars in the range. No longer does the outside front corner of the car dip towards the road if the car is flung into a bend (a tendency which was made worse, or became more noticeable with, the long-nosed Mark II cars). Now the nose stays nearly level, rolling no more than the back end, and the result is rather less understeer and great deal more predictability, not to say less spectacular progress.

However, for most testers it was not all good news, for the considerable price rises of the 1970s now made the 2500S look quite expensive. With an eye to profits rather than short-term praise, however, Rover-Triumph thought they could get away with this. In any case, development of the long-running range was now at an end, and the Mk II models only had another two years to run.

Rationalization had finally caught up with Triumph and Rover. In 1976 a new large hatchback, badged as a Rover, would take over, and the Mk II's career would be over.

10   1977: The End of the Road

When industry-watcher Jeff Daniels wrote a devastating appraisal of British Leyland's marketing strategy in 1980 (*British Leyland – The Truth About the Cars*, published by Osprey Publishing) he had this to say about the Triumph-Rover conflict:

> The resolution of the bigger-car question at Rover-Triumph was not easy. For some time after the British Leyland merger in 1968 it was assumed both inside and outside the Corporation that the Triumph name was assured of an inside track because it was in effect the vehicle which had taken Donald Stokes into the car industry and eventually into overall control of BL.
>
> It was an assumption which manifested itself most obviously on the sports car side, but it spilled over into the whole saloon-car area between the mass-market Austin-Morris range and the up-market Jaguars.
>
> There was no immediate reason why either the Rover or the Triumph 2000 should be chopped, least of all if it was likely to mean any drop in their highly profitable sales, but looking into the future it was not in BL's interest to develop two replacement models which would compete with one another in the Rover-Triumph tradition.

This was exactly what British Leyland's planners had already concluded: that it made little sense for this inter-marque fighting to continue indefinitely. By the early 1970s Triumph was looking ahead to what has been described as an outrageously ambitious programme ('over-extended' is just one quote I have), and the problems at the new Speke factory were taking up a huge amount of management time. In any case British Leyland was planning to rationalize its new-model programme, and Lord Stokes's right-hand man, John Barber, eventually wanted to see all its cars, of every marque, sold through all its showrooms. (You may ask whether it would ever have made sense to sell Jaguars alongside Minis, or Spitfires alongside Range Rovers, but at the time British Leyland's corporatist bosses seemed to be quite convinced on the subject.)

By the time the Rover-Triumph business had officially been set up in March 1972, the fate of the 2000 family was already sealed. Apart from a few wistful discussions at director level and a clay model or two, no serious efforts had ever been made to develop a replacement for the 2000 Mk II range.

Almost from 1968, it seems, Rover was assured that it would eventually get the chance of building a new medium/large model range, while Triumph was told of this, but assured that it would be left unchallenged to concentrate on smaller cars, more sporty cars – and sports cars.

> After working at Longbridge from 1969, on product planning [says Alan Edis] I moved fairly quickly to Triumph after that. There were some fairly big changes taking place within the organization. Bill Davis, who had been deputy managing director at Longbridge, moved to Triumph as managing director.
>
> He wanted me to go and establish Product Planning for him at Triumph. I set up an

office at Canley, I literally started on my own. I think Triumph's ambition was to have a niche as a small specialist producer.

At first we were never leaned upon by our elders and betters, not in terms of rationalization. But there was a picture which began to develop quite quickly, and before long we began to look at the project which became Rover's SD1.

At this time, if only Rover had been able to put its new large V8-engined P8 model into production (as a replacement for the old-fashioned 3.5-litre P5B models), there might still have been time for Triumph to persuade British Leyland's bosses that a slightly smaller Triumph should follow. After P8 was cancelled in 1971 (and a tooling investment of at least £5 million was written off), that slim chance disappeared for ever.

SD1 – A REPLACEMENT FOR THE 2000

The work that eventually led to a new Rover model – the SD1 hatchback – effectively taking over from the 2000 family, began in 1971. In the meantime Triumph had already started looking at ways of replacing the Innsbruck range with a car called the Puma and a shortened version called Bobcat.

According to Spen King, who was running the design team, it would have used the engines and transmissions that were later seen on other Rover-Triumph models 'with new suspension. We would have altered the rear suspension, with proper constant velocity joints – sorted-out versions of semi-trailing independent suspension.'

'These only got as far as small scale clay models,' Edis says, 'and I don't think we had even considered what components would have gone into them. There were good Triumph units available, six-cylinder and V8, but it never got as far as that.' After a brisk design competition had taken place

between Triumph and Rover, where scale-model proposals from Triumph (the Michelotti-inspired Puma) and Rover (the P10) were viewed at Rover's Solihull factory, the Rover project got the nod.

Rover was more advanced in its thinking, style and analysis of market needs than was Triumph. Soon after this head-to-head the P10 project evolved into the RT1 (RT = Rover-Triumph), before a new and definitive design called SD1 (SD = specialist division) was born in April 1971.

As a trade-off, though the situation was not quite as cut and dried as it sounds, Triumph was to produce a Dolomite-replacement coded SD2, a car in which Spen King's team put in a great deal of effort, though it was cancelled soon after the first prototype (X864 – SRW975N) had first run in 1975–6. Edis recalls:

A small group of us were beginning to look at the roots of the SD1 programme. The two organizations – Rover and Triumph – were still quite separate, but it clearly was not sensible for both Triumph and Rover to produce new cars.

It ended up as Spen King, David Bache [Rover's chief styling engineer], myself, Rex Marvin and Gordon Bashford [both of Rover], who began to work independently of what the organization was doing, and started to frame up the SD1. This happened in 1971.

After Spen King became engineering design supremo over Triumph *and* Rover in 1972, what followed was probably inevitable. As one of Rover's most noted designers, Gordon Bashford, once told me: 'Now we were expected to produce the best at a much lower cost. It was probably at this point that we finally decided to make the car technically much simpler. Among the changes was a decision to revert to a solid rear axle.'

As already mentioned, Nick Carver of British Leyland was moved from Berkeley

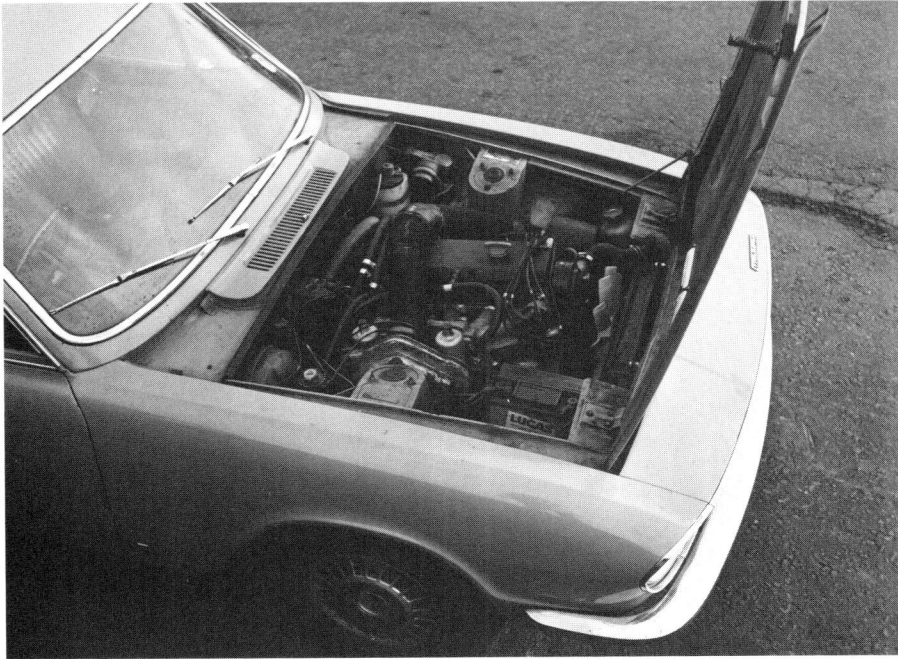

The one that got away! Triumph designed a new ohc six-cylinder engine in the 1970s, though in the end these 2.3-litre and 2.6-litre designs were only used in Rover-badged cars. Much of the test running, though, was carried out in late-model Innsbruck cars, which proves that this would have been an easy installation.

Square House in London to the Midlands, where his brief was to look after the product planning of the SD1 and all future Rover-Triumph models, and Alan Edis transferred his responsibility direct to him. His first priority was to define what SD1 was actually meant to be. To quote from my study of the Rover company, published some years ago:

> SD1, quite literally, had to be all things to all men. At Rover it would replace all existing models. The big old, but graceful 3.5-litre saloon would be allowed to die off (which it did, in May 1973), and the P6 family would be replaced by a family of SD1 cars. Eventually it would also have to replace the Triumph 2000/2500 cars, though these would carry on in production for a time at Canley after the first SD1s were built and sold.

As a postscript to the Triumph 2000 story, therefore, now is the time to detail some of the work that Triumph (more strictly, Rover-Triumph) engineers carried out for the new SD1.

A NEW OHC ENGINE

Well before BL chose to develop the SD1, Triumph had already spent some time, money and effort in developing a new breed of six-cylinder engine. In the beginning this new design had been proposed for use in Triumph cars, but well before its layout was settled, it was intended to power the six-cylinder versions of the SD1, which were badged as Rover 2300 and 2600. Strangely enough, I have never seen evidence of intent to use it in any other British Leyland car or light commercial vehicle – not even in the Stag where it would have offered an intriguing and equally powerful alternative to the troublesome V8.

As I wrote in a magazine article in 1978: 'Seven years elapsed from first thoughts to production engine, and the engine we see now is not the one Triumph sketched out in 1970. The engines were conceived by Triumph, altered by Triumph, developed by Triumph – and finally used in Rovers. What happened in 1977 is not what was planned in 1970.'

In 1970 Triumph was still functionally independent, and at this time technical director Spen King was being encouraged to devise a thoroughly modernized version of the existing 'six'. Jim Parkinson was in charge of the engine design department, but the actual design work was tackled by Mike Loasby (ex-Coventry Climax, who went on to achieve greater fame with Aston Martin and DeLorean in the 1970s and 1980s).

To retain as much of the original Triumph six-cylinder engine (and tooling facilities) as possible, the original intention was to carry out a straightforward overhead-cam conversion of the 2-litre engine, using the existing four main-bearing cylinder block. Technically this was not ideal, but a 'paper engine' was designed in 1970, which retained the existing camshaft (as a jackshaft to drive existing accessories), and had a chain-driven overhead camshaft in a new aluminium cylinder head, with in-line valves.

Nothing came of this until the Rover-Triumph merger was formalized in 1972, when Parkinson and Lewis Dawtrey were invited to try again, this time for an engine to be used in the SD1, but which could also fit into the existing 2000/2500 engine bay. However, any idea of re-engining the ageing Innsbruck model was abandoned: 'We expected to have the SD1 ready when the engine was ready,' Edis insists. 'We originally wanted to introduce the V8 version of the SD1 first, with the six-cylinder following on closely afterwards. In practice the development of the 6-cylinder engine took longer than we wanted . . .'

The logical course of action, therefore, was to let the existing Innsbruck model run on until the Rover 2300/2600s were ready, but not to let the two types overlap.

The revised design, coded PE146 (which meant Petrol Engine of 146 cubic inch capacity – 2.4 litres), was more ambitious. This time there was a modified cylinder block where the cylinder bore centres had been shuffled around to allow for an enlarged cylinder bore of 81mm, and a capacity of 2,350cc to take the place of *both* older engines; the traditional, age-old (1953-vintage) stroke of 76mm was retained. This time the overhead camshaft was to be driven by a cogged belt, and there was a two-valve/cylinder head with operation by the soon-to-be-acclaimed Dolomite Sprint system – one line of valves was directly operated, the other line by rockers. Using twin SU carburettors, more than 120bhp was immediately available.

Walter Hassan and Harry Mundy of Jaguar, plus Triumph's own Lewis Dawtrey, had all been involved in the concept work behind that novel layout. To summarize Mike Loasby's views: 'That was half-way better than a conventional single cam, and half-way towards a twin-cam without the complications.'

By 1973 British Leyland's planners had altered their requirements for the SD1 range yet again, this time demanding a larger engine and more power to help fill the gap between the PE146-engined Rover 2300 and the V8-engined Rover 3500. The 'Mk 2' engine design, therefore, featured yet another version of the cylinder block, cut down in height, and using Dolomite Sprint connecting rods. A longer stroke of 84mm helped produce a capacity of 2,597cc – and a full-blooded and torquey 136bhp. The mind boggles at the thought that a 2.9-litre version with an 86mm cylinder bore was also considered at this time.

Less than a year later the final design evolved. For the production-car layout the cylinder block was lengthened and strengthened, the cylinder centres were moved around yet again, and all traces of the original engine were finally eliminated. Originally intended to be unveiled in the Rover SD1 in 1976, it eventually made its bow in October 1977, and went into production a few weeks later.

Although Triumph's final engine was undoubtedly its best ever, its designers never got the praise that was due. During the development period Spen King and John Lloyd actually spent much time damping down their engineers' enthusiasm, to keep power and torque *down* to specified levels. It is now no secret that the 2.6-litre engine could easily have produced more than 150bhp with only minor changes to the specification.

As I wrote at the time, would the designers of the first overhead-cam engine, in 1970, ever have believed that the 1977–8 unit could look like it did? Now, too, I wonder if they would have been happy or disappointed with a design that powered only Rover 2300s and 2600s until January 1987?

A NEW GEARBOX

When the Rover 3500 made its debut in the summer of 1976, it featured a brand-new, five-speed, all-synchromesh gearbox. By the mid-1990s, descendants of that original design were still being used in cars like the TVR Griffith and Morgan Plus 8, and in fashionable 4x4s like the Land Rover Discovery

The other major new building block developed at Triumph in the early 1970s was a brand new robust, five-speed gearbox, which rendered unnecessary the use of an expensive Laycock overdrive. Originally fitted in the Rover SD1 hatchbacks, this box was later used in TR7s, Jaguar XJs, Morgans, TVRs and other specialist marques, and – in much modified form – in the four-wheel drive Range Rover.

and the Range Rover. In the 1980s, too, the same box found a home in the Triumph TR7 and TR8, and in some Jaguar saloons.

As with the engine just described, Rover took all the credit, but it was actually Triumph staff who designed it. Gearboxes, like axles (and on the new SD1 *that* was a Triumph design too), tend to do their work unseen and get no praise, but this unit, later named the LT77, was so classically simple, sturdy and well-nigh bomb-proof that it became well known. Any gearbox, after all, which can cope with the thunderous 325bhp power and 350lb/ft torque of the TVR Griffith 500 has a lot to be proud of. LT77, incidentally, means 'Leyland-Triumph 77mm', where 77mm was the spacing between the main and layshafts.

With future 'corporate building block' usage in mind, the new unit, complete with a neatly detailed remote control change, was meant to be versatile. Needed for the Triumph TR7, due to be launched in 1975, and already being considered for use in the Jaguar XJ saloons, it was designed by one of the most experienced teams in the industry. Even though George Jones had gone to Longbridge with Harry Webster, David Eley brought many years of experience to the job.

In layout, it had what is known as an 'overdrive' fifth gear. Fourth gear, in other words, gave direct drive from engine to propeller shaft, whereas fifth gear was geared up (the ratio was actually 0.833:1) to give longer cruising legs to any car which used it. As with most new British Leyland designs at the time, it was meant to do several different jobs, which is why there was also a four-speed version without the 'overdrive' fifth gear, which was only used in the Rover 2300.

Progress on the LT77 gearbox actually ran ahead of that for the overhead-camshaft six-cylinder engine, so much so that series production had begun before it was revealed for use in the Rover SD1/3500 model in July 1976. Soon after that it was made available on TR7 sports cars being sent to North America, and was later standardized on that sports car.

2000 AND 2500S – A GENTLE DECLINE

In spite of this intensive design work, however, there was never any intention to make the new engine, or the gearbox, available in the obsolescent Triumph 2000s and 2500s, though a lot of endurance running was done with the new hardware fitted into Triumph 2500TC/2500S cars.

Starting in 1973, with an overhead-cam engined 2500PI car (X829 – GWK336N), then with X833 (JRW695N), and X857 (GWK351N), a great deal of running was completed before true SD1 structures became available. In fact, the new engine did not go into series production until the winter of 1977–8, long after the last of the big Triumphs had been completed. By the time the Triumph 2500S had been introduced, British Leyland had already decided that no more development of the old cars would be carried out.

In spite of the financial traumas suffered by British Leyland in this period (which

included virtual collapse in January 1975, effective nationalization in the months that followed, and a horrifying loss of market share), the Triumph and Rover range was slimmed down markedly between 1975 and 1978. Here are the important events:

January 1975	Launch of new TR7 sports car
June 1975	**Launch of Triumph 2000TC/2500S**
July 1976	Last (six-cylinder engined) TR6 assembled
July 1976	Launch of new Rover 3500 (SD1) model
December 1976	Last Rover P6 assembled
May 1977	**Last Triumph 2000/2500 model assembled**
June 1977	Last Triumph Stag assembled
October 1977	Launch of Rover 2300/2600 models

In the summer of 1975 there had been seven 'executive' cars being built at Rover-Triumph – Triumph 2000TC, 2500TC, 2500S and Stag, Rover 2200SC, 2200TC, 3500 (and 3500S) – but by the summer of 1978 there were only two – Rover 2300/2600 and Rover

Year	Triumph 2000/2500	Rover 2200/3500/3.5 P5	Rover SD1
1973	21,000	23,627	–
1974	16,395	18,488	–
1975	12,333	15,280	–
1976	10,083	4,035	6,816
1977	6,307	4,813	12,374
1978	468	–	31,669
1979	–	–	29,576

3500. No doubt this made a lot of financial sense to British Leyland's planners, but it did not help Triumph in the showrooms.

British market sales (as opposed to production) in that period tell us as much about the decline of British Leyland as about the cars themselves. Far fewer of the new-fangled Rover SD1s were being sold in the UK in 1978 and 1979 than Triumph and Rover had sold between them in 1973 and 1974. Is there a lesson to be learned?

Over at Canley, assembly of the big Triumphs gradually slowed down. Before the energy crisis struck, 30,187 cars (no fewer than 12,934 with 2.5PI engines) were produced in 1973, but this was a high point. Production then tailed away, as shown in the table below.

Year	Total production
1973	30,187
1974	21,828
1975	17,838
1976	17,224
1977	8,370

After the last of the USA-market TR6s left the Canley assembly lines in July 1976, the big Triumphs were left as the only remaining users of the six-cylinder engine, though the four-speed gearbox (often with Laycock overdrive attached) continued to be shared with the Stag Grand Tourer until mid-1977. With the new Rover SD1 already in production at Solihull, and with the overhead-camshaft Triumph-engined Rover 2300/2600 due to join it in 1977, British

Leyland's planners began to run down the Triumph 2000/2500 lines at Canley.

The end came in May 1977 when the last of the line – three different types – were all built in quick succession: the last 2000TC was built on 17 May 1977, carried chassis no. ML29535DL, was a right-hand-drive carmine red saloon, and went to the home market; the last 2500TC was also built on 17 May 1977, carried chassis no. MM39373LDLO, was a left-hand-drive green saloon, and was exported to Holland; the last 2500S was built on 20 May 1977, carried chassis no. MP10253SCA, was a right-hand-drive green estate car with automatic transmission, and was delivered to the British Motor Industry Heritage Trust collection. It is now at Gaydon.

For Triumph, this was the end of an era. These were the last large cars ever to be built at Canley, and they brought to an end the sequence begun with the original Standard Vanguard of 1948. For Canley, and the assembly hall, this was yet another doleful step towards extinction. Although TR7 sports car assembly was transferred from Speke to Canley in 1978, no other new models were ever built there. The final run down began in 1980 when first the TR7, then the Spitfire, and finally the old Dolomite range, were all withdrawn. In November 1980 the lines were emptied, the lights were switched off, and the dust began to gather.

For the 2000 family, it was the end of a successful career. Between August 1963 and May 1977, every derivative of the range had been built at Canley. In that time, no fewer than 316,962 cars had been assembled and sold. No more elegant Triumph-badged cars were ever sold – and owners of classic 2000s, 2500s and PIs can be proud of that.

11 Stag, the Sporting Offshoot

As every Triumph enthusiast knows, the V8-engined Stag was a direct descendant of the Triumph 2000. Few realize, though, that the Stag which went on sale in 1970 was very different from the prototype conceived by Giovanni Michelotti in 1965. In a five-year gestation period, the car that started life as a convertible version of the 2000 became something unique in almost every way.

In the end the Stag had a relatively short and none too successful career. Launched in 1970 and discontinued in 1977, it sold only 25,939 copies. Its sales were not helped by the energy crisis of 1973, but it also suffered badly in North America. Great things had been expected of it there but in the end only 2,871 cars were sold. Federal-spec production ceased at the end of 1973, some cars remaining unsold until 1975.

The easy – but quite inaccurate – way to summarize the Stag's engineering is as a four-seater cabriolet/hard-top car with a short-wheelbase version of the 2000's under-pan, suspensions and transmission, but fitted with a V8 engine. In fact it was never quite all of those things at the same time. When the floorpan *was* a shortened version of the 2000 design, the V8 engine had not yet been adopted. The transmission became more and yet more specialized as development proceeded. Everything from wheels to brakes and to the basic structure changed persistently as time passed. Even so, by the time the Stag went on sale it was still possible to see where and how the connection with the 2000 could be made.

During this period, Triumph was working out its strategy for an entire new range of engines. After every possible combination had been considered, the company decided to develop single overhead camshaft units, slant-four-cylinder and closely related 90-degree V8, which could be used in many models. Even before it ran, Saab of Sweden placed a huge order for slant-fours, which ensured that deliveries would begin to them in 1968. V8 development and production had to lag behind, but the exciting new unit would be ready in 1970.

Like many other new Triumphs of the period, the Stag was not born in a product planning department, but in Michelotti's fertile brain. Finding himself at a loose end in 1965 (which, knowing the little Italian's hyperactive lifestyle, must have meant that he had half a day to spare!), he decided to build a special show car based on the Triumph 2000; it would be a convertible, still with four seats, though with reduced legroom.

Harry Webster, making one of his regular weekend dashes to Turin, remembers that Michelotti asked for a 2000 on which to base his ideas. Because most of the original car would be ditched, there was no need for him to be given a brand new machine. Instead, Harry decided to use this as an excuse to get rid of one of the old test cars.

6105KV, a very early car originally used as a model for publicity photographs, before being taken over by the engineering department, was used as a general purpose vehicle at the Le Mans 24-Hour race in 1965, then driven straight down to Turin, where Triumph lost sight of it for some months. By October 1965, almost everyone at Triumph had forgotten all about this car. Michelotti, however, had not. When Harry Webster walked into Michelotti's studio one day in February 1966, he saw the completed car:

Michelotti had requested an old Triumph 2000 on which he wanted to do a coupé/sports car based on the 2000 chassis, and he wanted to use it for a piece of advertising for his own skills. I agreed to this on the proviso that he didn't put it into a show until I had a look at it.

When I saw it I liked it, and thought this was a car that there was a need for, in Britain particularly, for the type of young manager who had some money to spare. It was the classic case of the car for a man who had a sports car when he married, then a family comes along. Soon he wants a car with a difference, with get up and go.

Conversely, there was the sort of man who, when the kids have left home and married, has quite a lot of money to spare, who still wants to feel young, and wants everyone to know about it. I don't know where this put me, but I thought I might want one of these myself. So I instantly made my decision on behalf of the company, and told Michelotti that I couldn't let him use it in a show now that he had finished it, but that I wanted to keep it!

In fact we then got it back to Coventry and kept it on ice for some time. It was the usual problem of priorities, and money to tool it.

Webster's decision needed guts, for only in 1963 had the idea of a convertible version of the 2000 been turned down flat by the sales force. In October they guessed that only 1,000 cars a year could be sold, and even though they revised that figure to 2,500 two months later, it still meant that no further work was done for some time.

By the summer of 1966 the first Michelotti show car/prototype was nearly complete. When Webster reported back to his board in July, he was very enthusiastic about the new model, which he christened 'Stag'. By this time Standard-Triumph was booming and profitable, demand for the Stag was estimated at up to to 12,000 units a year, and the project was approved. At that time it was to have a 2.5-litre, six-cylinder engine at first, with the new V8 engine to follow. Introduction was forecast for mid-1968, and the original estimate of tooling costs was £2.3 million. All this, of course, was pre-British-Leyland, a corporate change that would affect everything.

DEVELOPMENT HEADACHES

The very first prototype, converted from 6105KV, arrived in Coventry in June 1966, a conventional soft-top, centre-lock wire-spoke wheels and sliding headlamp shutters. At this stage it still had a 90bhp, 2-litre, six-cylinder engine, and Michelotti's own idea of what the facia/interior should look like.

By the spring of 1967, and still the only Stag in existence, this car had been given a stout roll-over bar, a hardtop (which would be optional), and a more powerful, six-cylinder fuel-injected engine. By that time Leyland had absorbed Rover, so at Sir Donald Stokes's request, attempts were then made to install the new Rover V8 engine, which had still not been publicly unveiled.

Although the directors were told that this engine could produce 160bhp 'which could be increased by using fuel injection', a few weeks later, to Stokes's consternation, Harry Webster reported to the board that there were difficulties because of the height of the Rover engine.

Spen King, who was closely involved with the Rover V8 engine from the day the company bought the rights from GM, told me:

This was the original Stag prototype, as built by Studio Michelotti, on the basis of an old, hack Triumph 2000 saloon. At this stage the headlamps were hidden by movable panels, and there was no suggestion of a roll-cage or T-bar.

The number plates on this, the second Triumph-built Stag prototype, are false. Originally assembled in 1968 with commission number X777, it had a six-cylinder PI engine. Eventually it was given a V8 engine transplant and became PVC237G.

191

Originally the V8 engine for Stag was intended to be a 2.5-litre unit with fuel injection, but in production form, as shown here, it was a 3-litre with two semi-downdraught Zenith-Stromberg carburettors.

I think they [Triumph] didn't want it to fit, they wanted their own V8 engine. I don't think they wanted the Rover to go in. Once I arrived, I didn't even have the option to try. But there was also a worry over the capacity of the Rover production lines. Clearly, in retrospect the wrong thing was done. The Stag could have been quite a successful vehicle if it had had a Rover V8 in it.

Then there was the scuttle shake problem. Even with stiffened floors and sills, early cars were by no means rigid enough. Still unsolved when Spen King took over from Harry Webster in 1968, it was eventually cured by a John Lloyd brainwave. Spen King:

It wasn't just noise, it was real movement. If you watched a piece of dirt on the screen when you hit a bump you could see the movement. It wasn't nice. John put in the connection between the roll over bar and the screen to make it a T-bar. I didn't think it would work, but it worked like a miracle. I don't think that had ever been done before.

Legend has it that the original experiments were carried out with a broom handle before the definitive bolt-in pressing was evolved!

In 1969, incidentally, a fastback coupé version of the Stag was proposed, and looked extremely promising. Spen King, by then a director of the Triumph division, reported his enthusiasm for the idea. In September 1969 the board agreed that sales of the fastback model might exceed those of the hard-top. The Michelotti style was not liked, though, and although Les Moore's rendering in 1971 was favoured, it was never put on sale.

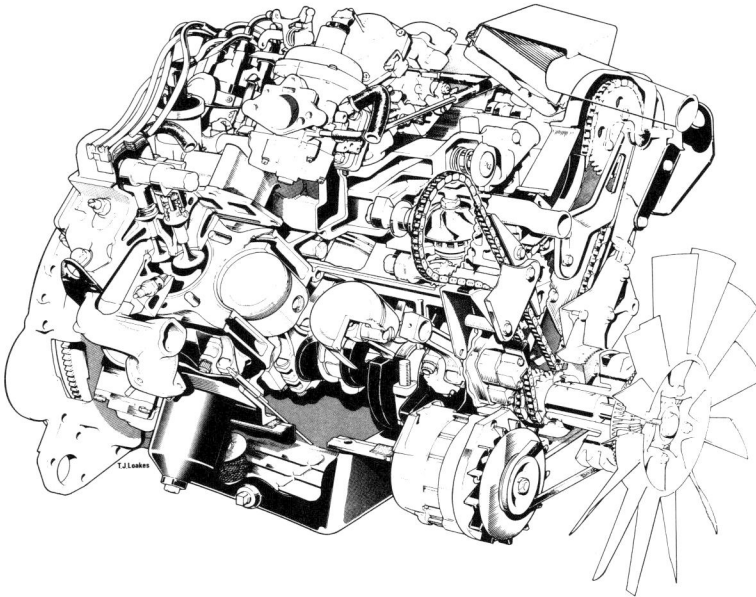

This cutaway drawing of the Stag V8 engine shows that it was a complex assembly. It shared many components with the Triumph slant-four engine used in Dolomites and TR7s, and supplied to Saab in Sweden. The V8 fitted easily into the Innsbruck saloon body shell, but no production car was ever made.

Choosing the Engine

Originally the plan was for the first Stags to have 2.5PI-type engines, and for the V8 engine to follow later, perhaps as an option. That is the way that Harry Webster certainly wanted it:

You can't change history, but my greatest wish was to use the 2.5-litre 'six': that was the way to go. That's the way it was always intended to come when I left it – and I would have left it that way. The V8 might have been right when properly developed, but when I left we'd hardly even run one, and we certainly hadn't put in the plant to make it.

Stag prototypes	–	1966	Converted from 2000 saloon 6105KV
The list of Stag prototypes, culled from Triumph's Prototype Register, confirms	X763	1967	With six-cylinder engine
that Stag development moved very slowly	X777	1968	PVC237G
at first. Although the first prototype was	X782	1968	PHP465G
running two years before the first Mk II	X783	1968	TKV754J
saloon was built, public launch was nine	X787	1968	
months after the launch of the saloon.		1969	(Body shell only, built up as X798)
Two years after the programme had officially begun only three prototypes had been	X790	1969	RHP659H
completed, all of which were initially fitted	X798	1970	
with six-cylinder engines.	X802	1969	
Here is the complete list:	X815	1971	WHP852J

> **Engineering: The 2000/2.5PI Connection**
>
> The Stag production car drew its inspiration from the platform of the 2000/2.5PI, but in many areas there were significant differences. Here is a summary of the connection.
>
> *Floor pan/platform:* Originally a short-wheelbase version of the 2000, many unique panels were eventually developed for the Stag. Many similarities remained in the front suspension/rear suspension/cross-members area.
>
> *Bodyshell:* Unique in almost every respect, though the grille style, details and front side/indicator lamps were shared.
>
> *Interior fittings:* Similar style and treatment, different in detail. Steering wheels shared, but instrument layout different.
>
> *Engine:* Unique, never used in any other Triumph, though it *could* have been fitted to the saloons. An ohc, 2,997cc 90-degree V8, with cast iron block and cast aluminium cylinder heads. 127bhp for USA-market cars, 145bhp for rest of the world.
>
> *Transmission:* Gearbox, clutch, overdrive, optional automatic transmission and final drive all basically the same, but upgraded in many details. These up-grades were later fed through to other models, notably the TR6.
>
> *Front suspension:* Basically the same as the saloons, but with an anti-roll bar, not seen on saloons (2500S only) until 1975.
>
> *Rear suspension:* Basically the same as the 2000/2.5PI.
>
> *Steering:* Basically as 2000/2.5PI, but power assistance as standard. Only optional on Mk IIs until 1975, then standard on the 2500S.
>
> *Brakes:* Bigger discs and wider rear drums than the Mk II saloons.
>
> *Wheels:* Wider-rim steel disc (later cast alloys) than 2000s. 14in Stag alloys (1973–7 models) were later standardized on 2500S, but never offered on other models. Centre-lock wire wheels also available – never offered on saloons.

The directors however, seemed determined to have their V8-engined Stag, for in September 1967 one of their meetings commented that the Stag could not be ready before 1969, and that: 'success depends on the success of the fuel-injected V8 engine'.

In TR5/TR6 tune the fuel-injected 'six' produced 150bhp (nett) – probably about 135bhp (DIN) – and this would certainly have given the Stag adequate performance, though low-speed torque was always a problem with that engine, as was the famously uneven idle which could only be tamed by detuning the unit.

To be honest, I suspect that the engine would have produced no more than 120/125bhp (DIN) when Stag-ready, and even then the fuel-injected engine could never have passed the ever-tightening USA emission regulations. For the USA, then, a six-cylinder Stag would have had no more than the 106bhp found in Federal-spec TR6 carbs of the period.

Spen King reckons that Harry Webster's gut feeling about the Stag was correct:

> There was a 2.5-litre version of the V8 running when I arrived, and it was very poor really. Low-speed torque of the 2.5-litre was very poor, and the carburation was still up the creek. The only way we thought we could sort it out was to go to 3 litres. We certainly

Gordon Birtwistle was testing X782, the third Triumph-built Stag prototype, at Mallory Park race circuit when this picture was taken. By this time the specification was almost settled.

had some cars with fuel injection – it must have been Lucas at the time, because Bosch wasn't available at that point.

But we were definitely told that the Stag had to use a V8, and that was it. The American side of British Leyland said it had to be a V8, with no question of doing anything else. The Rover V8 wasn't considered again, so we had to make our own V8 work.

One consequence of all this was that the transmission had to be beefed up with a larger clutch, a modified gearbox with slightly different ratios and bearings, and a stronger axle. Wheels went up to 14in and brake sizes were also increased, the result being that much commonality with the Innsbruck (particularly 2.5PI) model was lost.

Building the Cars

Although it was related to the Innsbruck models, the Stag took shape in a very different manner. Although final assembly was always at Canley, until 1976 the Stag was built on the TR6 assembly tracks, only moving to the Innsbruck line for the last year of its life.

During the design/development phase, the Stag shell had moved progressively away from the 2000's layout, as John Lloyd confirms:

We thought we could get away with modified 2000 tooling parts, but by the time we had had to chop off a bit here, add on a bit there, and particularly to provide roll-over protection, there was precious little left of the original pressings. Our body design people than pointed out that all-new tooling would be as cheap as modified tooling, especially if we did most of it ourselves.

Accordingly, most body panels were pressed in two sites at Speke, Liverpool – in the

The Stag production car went on sale in 1970, at that time easily the most complex Triumph car ever offered for sale. The T-bar was a permanent fixture, though it could be unbolted for maintenance and repair purposes. The hard-top was optional.

'Liverpool No. 1' factory (the centre of 1300/Dolomite/TR6 shell manufacture) and at the vast new 'No. 2' factory where TR7s would later be built in their entirety. Shells were assembled, painted and trimmed in Liverpool before they were transported to Canley for the final assembly process. Engines, transmissions and axles were all manufactured in Coventry, either at Canley or Radford.

STAG ON SALE

Although the Stag was launched in June 1970, sales did not actually begin until the autumn, and it was already clear that the specially modified federal version (with detoxed engine, and with safety features built into the body shell) for sale in the USA would not be ready until the autumn of 1971. Even then, many cautious personalities at Triumph thought the Stag was still not ready

The Stag's facia/instrument display was similar in layout to that of the 2.5PI, though there were many detail differences.

Viewed from above, the rather restricted four-seater layout of the Stag is obvious. Getting into the rear seats without banging one's head on the T-bar was not easy.

to sell in the USA, but British Leyland's sales force insisted that it should be made available.

By the time deliveries began in numbers, Stag prices started at £2,156 (which, for comparison, was £378, or 21 per cent, more than that of the 2.5PI Mk II saloon of the day). Waiting lists built up at first, but by 1972 production balanced sales fairly accurately. Deliveries to the USA began before the end of 1971, when the lowest quoted price (for east coast delivery) was $5,525, the equivalent of £2,302. When all the desirable extras were added, that price rose by about $1,000/£417.

For the next six years Triumph fought a grim battle to get, and keep, the engines reliable, for there was an early history of cooling problems and cylinder headgasket failures, too widespread to be hushed up. In the early days, too, USA sales suffered because the Stag was thought to be expensive, yet neither as fast nor as fuel-efficient as originally hoped; this was because the exhaust

The Stag's soft-top folded neatly back into a recess behind the seats, and was then covered by a fold-down panel.

197

Stag styling was retouched for 1973, though there were no changes to the sheet metal or the shape of the 2+2-seater layout.

Stag production

The very first off-tools Stag was built in November 1969. Only 114 such cars had been built before the car was announced in June 1970. Peak production came early, in June 1972, when 589 cars were assembled. The last Stag of all was assembled in June 1977. Annual production was as follows:

1970		740
1971		3,901
1972		4,504
1973		5,508
1974		3,442
1975		2,898
1976		3,110
1977		1,836
Total	Home Market	17,819
	Export	8,120
	Grand total	25,939

emission regulations were strangling performance, a problem encountered by many other European manufacturers at this time.

Triumph revised the Stag for 1973, this version now familiarly being known as the 'Mk II' though Triumph never called it that. Style changes included the use of cast alloy road wheels as standard on federal-spec cars, while overdrive was standard on manual transmission cars. USA-market production, however, ended after 1973, when Triumph was no longer willing to keep abreast of burgeoning USA safety legislation (including the use of '5mph' bumpers, which had to be distortion-free in shunts up to that speed (7.5km/h)).

The energy crisis and petrol price rises of 1973–4 had a big effect on all cars like the Stag, which had large engines and which were seen (rightly or wrongly) as self-indulgent machines. Thereafter Stag sales

There was a serious proposal to produce a fastback hard-top version of the Stag, which only fell down because the investment needed to tool for body changes was too high. More practical than the removable-hard-top Stag which did *go on sale? Make up your own minds . . .*

(particularly to export markets) sagged gradually away, so much so that plans to replace the car with a long-wheelbase TR8, coded Lynx were cancelled.

Although Rover V8 engines were fitted to a few Stag development cars in the mid-1970s nothing was done to turn that into a production package. 1976 and 1977-model Stags had cast alloy wheels as standard, along with tinted glass and a slightly different cosmetic package, while Borg Warner Type 65 automatic transmission took over Type 35 transmission as an option on the last year's production.

Once the 2000TC/2500TC/2500S cars had dropped out of production in May 1977, it was inevitable that the Stag would follow. The very last one of all was produced on 24 June 1977, bringing the grand total to 25,939 cars.

Appendix A

BUYING AND RESTORATION

Although the last of these Triumphs was built nearly twenty years before this book was first published, a fair proportion of them seem to have survived. By the 1990s these cars, especially the 2500S and all injected types, were accepted as classics, well worthy of preservation.

It was interesting to note that although the British Leyland model that displaced the Innsbruck range – the Rover SD1 – was virtually ignored by latter-day enthusiasts, many extremely smart examples of the big Triumphs were totally restored and used by members of the one-make clubs.

In many ways, all these cars are classics because of their elegant good looks, their tasteful and high-quality furnishings and their comprehensive equipment. There is no doubt in my mind, though, that the best *and* the worst features of this type are found in the 2.5PI. There was never any basic problem with the chassis or the drive line – most problems arose in the Lucas fuel-injection system. Such an installation, in good adjustment, gives a wonderfully smooth engine with a lot of very real power; a bad one can not only give poor performance but poor fuel consumption as well.

On balance, therefore, a late-model 2500S is probably the best buy, because it is little slower than a 2.5PI but much more mechanically simple, and possesses the best handling of all types.

As to styling, most people agree that the Mk II cars looked better, and everyone agrees that the Mk II facia/instrument layout was superb for its day. Although they were quite expensive, the estate cars were extremely successful, with all the style and comfort of the saloons, along with a very useful (if not vast) loading volume, fully trimmed like the rest of the car. If you must have a modern 'classic estate' to act as a tender car to your other classics, then a big Triumph estate ought to be ideal.

In any case, because of the large numbers built, there should still be a wide choice of all types – saloon or estate, manual or (more rare) automatic, 2-litre or 2.5-litre engine. Even so, the cars were not perfect, by any means, particularly because they were designed and manufactured (at first) by a company not exactly flush with profits, so anyone looking to buy one of these cars in the 1990s should look out for the following characteristics.

Body Structure

These were always big, solid cars and even those in quite an advanced state of decay can often still sail through the compulsory MOT test, particularly if the underside of the structure is still sound. However, always remember that an early Mk I would now be around thirty years old, and not even a Rolls-Royce lasts that long before needing expensive attention!

The major corrosion areas in these cars include the inner and outer sills (which contribute a good deal to the structural beam strength of the body), the front wings and the wheel-arches, the bottom edges of the doors and the front corners (near the side lamp/indicator flasher assemblies) and the pick-up points for the front bumpers.

Structurally, too, it is very important to see if the top suspension strut mountings (visible inside the engine bay) are sound, for they can rust through from the outside, and to look at the condition of the pick-up points for the rear suspension subframe at each side of the car under the rear seat.

If there is any water leaking into the front seat footwells or into the boot, then you may be in for expensive repair work, for corrosion in the floor panels, boot floor, or even the front toeboard will be the cause of this.

With this body there should be no question of doors having dropped, but in case the cars have been in major accidents during their lives it is always wise to check for the standard of fit and finish of the two really large external panels, the boot and the estate car hatch. The bonnet, in particular, is vulnerable to accident misalignment because its hinges are at the front of the car.

Suspension and Steering

If the body shell that supports all the suspension mountings is in good shape, the suspension should be sound, but before buying a car it is essential to ascertain that the geometry (*especially* at the rear) is right, for this not only affects straight-line running, but the rate of tyre wear as well.

Damping on these cars is generally quite soft (even softer on the final examples that have front anti-roll bars) but on the estate cars the rear dampers in particular have a hard time, and may need replacing. At least they are not desperately expensive and are simple to replace. At the front, McPherson struts look more complex than they are, and replacement damper inserts can be bought to bring them back to health.

Depending on the age and particular model in this range, there are several different rates of front and rear spring – estate car rates differ from those of saloons, for example – and by now, you may find that some cars have been 'restored' using incorrect-spec components. Have a look at the parts manual if you have any doubts, and do not be too ashamed to take advice.

There was no front anti-roll bar on any 2000 until the 1975–7 period, but if you are not too worried about total originality, this addition could firm-up the handling of an earlier model; this will involve some fiddling around with brackets and bushes, but is not too complex a modification to achieve.

If the various suspension ball joints and bushes have worn, the handling will feel particularly loose, and not at all precise. The steering could be heavy, particularly if mounting rubbers have deteriorated and the alignment is no longer within limits.

Power-assisted steering was optional on later models, but standard on the 2500S (for details see the specifications for each model), and is now considered to be very desirable. Unhappily, upgrading manual cars to power cars is not at all easy, as a lot of necessary extra engine kit – pumps, belts, pulleys and other details – adds to the cost of buying a rack, pipes, reservoirs and other details.

By now I hope that all surviving 2000s and 2500s will be running on radial-ply tyres, which are much to be preferred. It is false economy to fit cross-ply tyres (if, even, you can find them these days) merely because they were original equipment when 'your' car was new. In addition, please do not 'over-tyre' your basic 2000, which really does not need the 185-section rubber of the more powerful derivatives.

Engines, Transmissions and Brakes

The basic six-cylinder engine is very rugged indeed, and apart from having to work at keeping the twin-carburettor linkage and settings in good shape, is very simple to keep in good condition. The major problem, which actually caused the 2.5PI to be dropped before its time, was the Lucas fuel-injection system, which was unreliable, difficult to keep in tune, expensive to restore and on badly kept examples often resulted in high fuel consumption and poor performance.

Fortunately, the application of money to the problem, and consultation of specialist firms now expert in this design (not Lucas, who lost interest many years ago), means that all can be put right, even today. A healthy 2.5PI is a joy to drive, and at least all the parts (remanufactured, repaired or redesigned) are available. Beware any PI cars with persistent fuel smells from under the bonnet, high-speed misfires, or generally uneven running. This will never clear itself, but will only get worse, and it might cost a great deal of money to eliminate. Replacement distributor units and pumps are very expensive.

The engines themselves eventually show their age with heavy oil consumption, noisy valve gear and a generally oil-leaky appearance. The blue smoke in the exhaust syndrome is one to fear, as is noise from the timing chain or the water pump. Exhaust manifolds may have cracked, giving the engine a generally uncouth sound.

Even the very best large Triumph gearboxes have a notchy feel, but this was always a characteristic of the design, caused by the detail arrangement of the very strong synchromesh. In rare cases of worn synchromesh, rebuilds are easy enough and all parts are still available.

Overdrive, by no means fitted to all cars (though optional right from the start) is a very desirable fitting, as without it these cars tend to feel a bit under-geared and fussy. The usual Laycock overdrive faults should be investigated when you go to inspect a car – sluggish engagement and jerky oper-

ation – both of which can indicate either dirty lubricant or badly worn friction surfaces.

Overdrive trouble, however, tends to be electrical, for the solenoid is exposed to attack by water, dirt and flying objects, so if the overdrive does not work on one of these cars, it may not be as expensive as all that to make it operative again.

There is little to report about the brakes, except to point out that in my opinion they are not really large enough to cope with all the performance of the 2500S and 2.5PI models; originally the front discs had to be squeezed inside 13in road wheels, which did not help, and of course the rear brakes were drums. Standard-specification pads are relatively soft, to improve the feel and to eliminate squeal, but this does mean that they wear out quite frequently, particularly if a car is driven hard.

At the rear, the exposed drive shafts tend to dry up their sliding joints with age; since the whole operation of the rear suspension depends on these joints actually continuing to slide freely, you should always be sure that they have not seized, or cornering can be a very jerky and uncomfortable business. Make sure, too, that the rubber boots around the sliding joints are intact.

Interiors and Decoration

Generally speaking, the trim and furnishing of 2000s was of high quality. However, my opinion is that the early facia/instrument styles, with their black-on-white dial instrument designs, were not as smart as the later types (some cars, in fact, have been updated by later owners, as our pictures make clear), and of course the Mark II (long-nose body shell) versions had the nicest and best equipped facias of all.

Although trim items for these cars have been remanufactured, it is always wise to keep the original panels if they have not worn badly. Accordingly, if you are intent on

Almost everything a Triumph owner might need for a major rebuild: this display at a one-make car meeting includes complete cylinder heads, radiators, springs, bumpers, badges and electrical items.

buying such a car, be sure that the furnishings, particularly soft trim panels and seat covers, are in good and complete condition.

Although the estate car versions were all sumptuously trimmed, the loading areas were particularly vulnerable to damage by bulky objects clumsily loaded or stacked in the car. In particular, look at the state of the wood cappings on the doors, close to this danger area. Do not worry too much if the carpet on these cars is damaged, as such items can now be remanufactured with a great degree of accuracy and authenticity. If you are looking for an estate car, incidentally, do not forget to check that the seat folding and latching arrangements work properly, and

that the tailgate not only works but keeps out the water. In such areas, a good clean up and lubrication of all moving parts usually help, and renewal of seals around boot lid and hatchback extremities often works wonders.

Although many parts are no longer available for these cars (body panels and new major castings, for example), it is usually possible to put a reasonably maintained 2000, 2500 or 2.5PI back into good condition if you are determined to do so.

I certainly recommend that you should join the one-make club covering these cars, and you should also try to get hold of authentic parts manuals and service manuals to help you maintain the car properly.

Appendix B

TRIUMPH 2000/2500 FAMILY PRODUCTION FIGURES

Mk I	2000 Saloon	2000 Estate	2.5 PI Saloon	2.5 PI Estate					Annual totals
1963	1,685								1,685
1964	18,490								18,490
1965	18,880	207							19,087
1966	21,237	2,194							23,431
1967	18,148	1,672							19,820
1968	19,264	1,824	1,154						22,242
1969	10,506	1,135	5,365	223					17,229
Total	108,210	7,032	6,519	223					121,984

Mk II	2000 Saloon	2000 Estate	2.5 PI Saloon	2.5 PI Estate	2500 TC Saloon	2500 TC Estate	2500 S Saloon	2500 S Estate	Annual totals
1969	4,950	459	2,139	148					7,696
1970	20,647	1,337	9,323	767					32,074
1971	19,475	1,520	9,896	801					31,692
1972	16,582	1,621	8,702	1,164					28,069
1973	15,904	1,349	11,899	1,035					30,187
1974	7,788	970	3,117	335	9,314	304			21,828
1975	5,749	321	416		8,852	369	1,119	1,012	17,838
1976	4,250				8,635		3,423	916	17,224
1977	1,658				5,018		1,391	303	8,370
Total	97,003	7,577	45,492	4,250	31,819	673	5,933	2,231	194,978

Index